FARMING *in* NATURE'S IMAGE

ABOUT ISLAND PRESS

Island Press, a nonprofit organization, publishes, markets, and distributes the most advanced thinking on the conservation of our natural resources—books about soil, land, water, forests, wildlife, and hazardous and toxic wastes. These books are practical tools used by public officials, business and industry leaders, natural resource managers, and concerned citizens working to solve both local and global resource problems.

Founded in 1978, Island Press reorganized in 1984 to meet the increasing demand for substantive books on all resource-related issues. Island Press publishes and distributes under its own imprint and offers these services to other nonprofit organizations.

Support for Island Press is provided by Apple Computer, Inc., Geraldine R. Dodge Foundation, The Energy Foundation, The Charles Engelhard Foundation, The Ford Foundation, Glen Eagles Foundation, The George Gund Foundation, William and Flora Hewlett Foundation, The Joyce Foundation, The John D. and Catherine T. MacArthur Foundation, the Andrew W. Mellon Foundation, The Joyce Mertz-Gilmore Foundation, The New-Land Foundation, The J. N. Pew, Jr., Charitable Trust, Alida Rockefeller, The Rockefeller Brothers Fund, The Rockefeller Foundation, The Florence and John Schumann Foundation, The Tides Foundation, and individual donors.

FARMING *in* NATURE'S IMAGE

An Ecological Approach to Agriculture

Judith D. Soule and Jon K. Piper

Foreword by Wes Jackson

ISLAND PRESS

Washington, D.C. ■ *Covelo, California*

The authors are grateful for permission to include the following previously copyrighted material: Excerpts from *Agroecology,* edited by C. R. Carroll et al. Copyright © 1990 by McGraw-Hill, Inc. Reprinted by permission of the publisher, McGraw-Hill, Inc.

Library of Congress Cataloging-in-Publication Data
Soule, Judith D.
    Farming in nature's image : an ecological approach to agriculture / Judith D. Soule and Jon K. Piper.
        p. cm.
    Includes bibliographical references and index.
    ISBN 0-933280-89-0 (cloth).—ISBN 0-933280-88-2 (paper)
    1. Sustainable agriculture—United States. 2. Agricultural ecology—United States. I. Piper, Jon K. II. Title.
S441.S757 1991
630—dc20   91-21120
                                                                        CIP

Printed on recycled, acid-free paper

Manufactured in the United States of America
10 9 8 7 6 5 4 3 2

We dedicate this book to our children:
Jacob, Bethany, and Erica Soule
Joshua, Emily, and Samuel Piper

Contents

Foreword

THE CHECKLIST OF environmental and social costs that must be charged against industrialized agriculture is increasingly well known even as the list expands. This book goes so far beyond that checklist into an analysis sufficiently sobering that the reader soon realizes that real solutions are less in the realm of "smart resource management" and more in the change of our world view—in short, a change in the way we see nature. If the world is a mine, then "smart resource management" is a logical answer. If the world is a source of hope, as the ancients believed, then a research agenda based on the way the world has worked for millions of years becomes logical.

With this volume, then, we finally have the most thorough-going argument for and meaning of "farming in nature's image." This is true because Judy Soule and Jon Piper have bothered also to supply the necessary details of what this means. It is the most sensible approach to farming because it is the most necessary approach if we are to continue to use agricultural land for as long into the future as we have had agriculture in the past. We are at that exact point in history where we must make some crucial decisions. The "problem of agriculture" has been exacerbated by modern industrial agriculture. These professional ecologists bring their rigorous scientific training as well as their ecological perspective on sustainability and convincingly tell us how to proceed as they make the case for the possibility of a far less wasteful way to farm. They are writing about nothing less than a marriage of ecology and agriculture.

Tangible research results provide the tone of authenticity we experience here. And not just the results of other people's work either, for the prescriptions resonate from the authors' firsthand experience with on-the-ground experiments that have yielded and that continue to yield data from a research agenda in which nature is the standard or measure. When the dentist taps a tooth, he or she is listening to the tone in order to learn how grounded it is in the bone. Having written these prescriptions "grounded in the bone," Soule and Piper are part of a small but emerging breed of young scientists who recognize that there are countless social, political, and cultural barriers that stand in the way of carrying out the vision of which they write. One hopes and expects that the classic papers for a sustainable agriculture are now in the making.

In 1978, I published a small paper entitled "Toward a Sustainable Agriculture" in which I argued for an agriculture based on the way the prairie works. I later expanded and more fully developed the argument for a natural solution to the "problem of agriculture" in a small volume entitled *New Roots for Agriculture*, which was published in 1980. The book in hand has built on that "nature as standard" notion. Soule and Piper's thinking about an ecological agriculture has gone far beyond where we were ten years ago. Not only have they rephrased and improved the relevant scientific questions I had posed for such an agriculture, they have broken those questions down to more manageable portions, a necessity if we are to more precisely contemplate a workable research agenda over the long pull.

It may sound like the idea began with my 1978 and 1980 publications. Nothing could be further from the truth, for this idea of "nature as standard" or "nature as the measure," as Wendell Berry has pointed out, goes back at least 2000 years before Jesus of Nazareth. In a memorable speech delivered at the dedication of our new greenhouse at The Land Institute in 1988, Wendell Berry traced the literary and scientific history of our work. He began by citing Job, who said:

> . . . ask now the beasts, and they shall teach thee; and the fowls of the air, and they shall tell thee:
> Or speak to the earth, and it shall teach thee; and the fishes of the sea shall declare unto thee.

Next he mentioned Virgil, who, at the beginning of *The Georgics* (36–29 B.C.), advised that

> . . . before we plow an unfamiliar patch
> It is well to be informed about the winds,
> About the variations in the sky,
> The native traits and habits of the place,
> What each locale permits, and what denies.

He then moved on to describe the writings of Edmund Spenser, who toward the end of the 1500s called nature "the equall mother" of all creatures, who "knittest each to each, as brother unto brother." Spenser also saw nature as the instructor of creatures and the ultimate earthly judge of their behavior. Shakespeare, in *As You Like It*, has the forest in the role of teacher and judge. Milton, in *Comus*, has the lady say:

> she, good cateress,
> Means her provision only to the good
> That live according to her sober laws
> And holy dictate of spare Temperance . . .

And, finally, Alexander Pope, in his *Epistle to Burlington*, counseled gardeners to "let Nature never be forgot" and "Consult the Genius of the Place in all."

Wendell Berry says that this theme departs from English poetry after Pope, with the later poets seeing nature and humans radically divided. A *practical* harmony between land and the people was not on their agenda. Even the Romantic poets placed such preeminence on the human mind that nature wasn't anything to deal with in a practical sense so much as what Wendell Berry referred to as a "reservoir of symbols."

We have largely ignored this literary tradition. For what if the settlers and children of settlers whose plowing of the Great Plains in the teens and twenties gave us the dust bowl of the 1930s had heeded Virgil's admonition "before we plow an unfamiliar patch / It is well to be informed about the winds"? What of Milton's insight about the good cateress who "Means her provision only to the good / That live according to her sober laws / And holy dictate of spare Temperance"? Virgil was writing about agricultural *practices* while Milton was writing of the spare use of nature's fruits.

But much of this book is about the *science* of agricultural sustainability where nature is the measure. Are there any *scientists* who have spoken in the same manner as the poets? In his dedication speech,

Wendell Berry went on to point out that after having gone underground among the poets, when the theme surfaced again it was among the agricultural writers who had a scientific bent. Liberty Hyde Bailey's *The Outlook to Nature* appeared in 1905. The grand old Cornell University dean described nature as "the norm": "If nature is the norm, then the necessity for correcting and amending abuses of civilization becomes baldly apparent by very contrast." He continued: "The return to nature affords the very means of acquiring the incentive and energy for ambitious and constructive work of a high order." It certainly does, and Soule and Piper illustrate that well. Later (1915), Bailey's *The Holy Earth* was published. In it Bailey advanced the notion that "a good part of agriculture is to learn how to adapt one's work to nature . . . To live in right relation with his natural conditions is one of the first lessons that a wise farmer or any other wise man learns." True enough, and now Soule and Piper begin to spell out a research agenda.

Sir Albert Howard published *An Agriculture Testament* in 1940. Howard thought we should farm like the forest, for nature is "the supreme farmer." He wrote:

> The main characteristic of Nature's farming can therefore be summed up in a few words. Mother earth never attempts to farm without live stock; she always raises mixed crops; great pains are taken to preserve the soil and to prevent erosion; the mixed vegetable and animal wastes are converted into humus; there is no waste; the processes of growth and the processes of decay balance one another; ample provision is made to maintain large reserves of fertility; the greatest care is taken to store the rainfall; both plants and animals are left to protect themselves against disease.

Before Howard's later writings, in 1929, J. Russell Smith's *Tree Crops* was published. He too believed that "farming should fit the land." Smith was disturbed with the destruction of the hills because "man has carried to the hills the agriculture of the flat plain." An agriculture modeled on the prairie featuring perennials would make hillside harvest of seeds possible.

Now we come to a most important point about the volume at hand. It may appear that it is part of a succession in a literary and scientific tradition that places nature as the measure. But as Wendell Berry said about the poets and scientists quoted above, yes, there is a succession, but the understanding probably comes out of the familial and communal handing down of the *agrarian common culture,* rather than any succession of teachers and students in the literary culture or in the schools. These literary and scientific examples, in other words, have *emerged out of* the common culture. The writers did not build on other writers who had gone before. Therefore, as Wendell Berry says, they form a series, not a succession.

Those who popped out of that common culture to form that series—both poets and scientists alike—have made us their successors. Not quite. What we can hope for with this volume is that it will be regarded in fifty years or so as one of the early stages of a succession, because finally we have something to build on in a scientific sense. It gets down to experiments and data. Both are here, though they are not the book's dominant features. In other words, besides the ideas, there is the sort of content a scientist can get his or her teeth into. Even though this volume shows the practical possibility of a research agenda based on a marriage of agriculture and ecology, it will require a push from those who have examined the assumptions of modern agriculture versus what nature has to offer and decided in favor of learning from nature's wisdom.

I emphasize this because the research we are talking about as necessary won't just "happen." There is no guarantee that fifty or a hundred years from now this volume will be viewed as an early stage of a scientific succession. The time is shorter than most of us realize. Most of the scientific knowledge of the past that makes this research (and, for that matter, essentially all research) possible is based on the slack that both fossil fuel and a young continent afforded. Such slack is disappearing fast. The fuel is running low, national ecosystems are in decline everywhere, and one billion people seem likely to be added to this planet in the 1990s alone! Nevertheless, if this book is taken seriously and young scientists continue to build on this greatly advanced perception of using nature as a measure, perhaps this Johnny-come-lately thing we call industrial agriculture will have been a mere blip, something we overcame because we came to our senses in time. Many of us are hopeful that industrialized agriculture does end up being an anomaly that came somewhere *in the middle* or toward the beginning of the human experiment with agriculture rather than *at the end*. We would like for our optimism to match that hope.

Wes Jackson,
The Land Institute, Salina, Kansas

Preface

THE IDEA FOR this book can be traced to The Land Institute's "Sustainable Agriculture Curriculum," an exhaustive, many-paged outline and stack of five looseleaf notebooks containing seminal papers in such fields as population genetics, crop breeding, ecosystem science, and prairie ecology compiled by Marty Bender and Wes Jackson in the early 1980s. This stack was handed over to us unceremoniously upon our hire as successive Land Institute research associates, and we were asked to teach the contents, or something resembling that. Part of the original motivation for this book, then, was to produce, in a more portable text, a synthesis of ecology and agriculture to be used in classes at The Land Institute and elsewhere. As we launched into the project, however, we soon veered from the textbook format. We decided that what we wanted to write was not a textbook-style amalgamation of traditional ecology and agriculture, but a scientifically backed plea for the radically new form of agriculture first laid out in Wes Jackson's *New Roots for Agriculture* and now being researched at The Land Institute. The purpose of our book is to demonstrate that the economic problems of farmers and rural communities, and the environmental damage resulting from modern agricultural practices, have roots in the industrialization of agriculture. We hope that our treatment here will bring renewed and serious attention to the importance of a new approach to solving agricultural problems, an approach that draws on models provided by natural ecosystems, to redesign agriculture in nature's image.

To a large extent, it is hard for us to claim this book exclusively as our own because the ideas herein so reflect the influence of many people. Much of what we have written has been brought up for consideration in the lively warm-up discussions that are an almost daily part of The Land Institute. We thank all past and present members of The Land Institute for their contributions. We are grateful to Wes Jackson and Dana Jackson for their challenge, support, and inspiration for as long as we have known them and for putting so much of themselves into making The Land Institute a place that inspires so many. Wes helped write the original book outline, directed us to the publisher, reviewed the first and last rounds of drafts, and provided gentle guidance, but kept a distance that allowed this book to be our own creation. Dana contributed ideas and references along the way and also read the last draft of Chapters 2 and 6. We thank Chuck Francis and Jack Ewel for their encouraging and frank comments on the entire manuscript and Phil Robertson for his review of Chapter 1. Chuck Francis and Peter Kulakow provided comments and discussion for the section on crop breeding in Chapter 2. Danielle Carré helped devise the original outline that eventually became this book. Some of her words show up in Chapter 1, where we borrowed from "Ecological Consequences of Modern Agriculture," coauthored by Judy, Danielle, and Wes, and published as a chapter in *Agroecology*. Many other colleagues, too numerous to name here, provided references, conversation, and information along the way. Bob Soule tirelessly edited many portions of many drafts. The patience and continual enthusiasm for the project shown by Barbara Dean, our editor at Island Press, kept us going even when deadlines passed and it seemed the project would never be completed. She also helped improve organization and tone during the development of the manuscript. Barbara Youngblood, also of Island Press, was a great help with the finishing touches.

Island Press provided Judy with advance funding, which allowed her to arrange enough free time to produce her portions of the initial draft. The Land Institute supported Jon for phone calls, photocopying, and some of his time.

Most importantly, a project like this can never happen without a great deal of moral support. Our spouses, Bob and Beth, deserve a large portion of the credit for this book, because without them we would never have finished even a first draft. We thank them for their faith that we could all survive the process, for crucial bouquets, chocolate, and dinners out, for unending encouragement and confidence, and, especially, for their willingness to sacrifice their personal time to take over our home responsibilities for weeks at a time as we pushed through various deadlines. Finally, we thank our children for reminding us daily of the importance of working toward a sustainable future.

FARMING *in* NATURE'S IMAGE

Introduction

ALONGSIDE INTERSTATE 70 in Kansas, between Manhattan and Salina, stands a rather small billboard with a picture of sunshine and wheat and this proclamation of state pride: "One Kansas farmer feeds 96 Americans, and you!" This simple statement has profound implications for both the economic and ecological status of modern agriculture. By the measure of productivity expressed in this sign, modern agriculture is tremendously successful; yet it is of growing concern whether this productivity is sustainable, either ecologically or economically. How much longer can society afford to pay the ecological price of this productivity: eroded soil, polluted groundwater, and pesticide-contaminated workers, soil, and food? How much farther is the nation willing to go on the road toward concentrated ownership of farm assets? Are citizens willing to close the door on the next generation of family farmers? The 1980s were critical in drawing the nation's attention to the vulnerable economic structure and crumbling ecological support systems that underlie modern agriculture. As evidence that these concerns are widely held, 60 organizations from across the country signed a joint letter to President Bush calling for a "full-fledged, across-the-board" effort at the U.S. Department of Agriculture (USDA) to develop the potential of low-input alternatives in agriculture. The 1990 Farm Bill addressed some of these concerns by extending the Conservation Reserve Program (CRP), changing subsidies to provide opportunities for farmers to use rotations without losing benefits, and setting national standards for "organically grown" labeling. From environmental groups such as the National Audubon Society and the Natural Resources Defense Council to grassroots farmer organizations and rural advocacy groups such as the Land Stewardship Project and the Center for Rural Affairs, and even to the land grant agricultural schools, debate is focusing on how to make farming more ecologically and economically sustainable.

While the economics and ecology of agriculture are profoundly intertwined, a sound ecological basis is essential for the long-term sustainability of agriculture, simply because agriculture is essentially and primarily a biological system. Yet the immediacy of economic crises often clouds the nation's vision of the ecological consequences of economic choices. In the 1980s the economic and ecological problems of agriculture reached crisis proportions simultaneously. This coincidence helped to spotlight their interrelatedness. Although this book deals primarily with the ecological aspects of agriculture, it is crucial to see the connections between ecological, economic, and social consequences of modern agriculture. Following is a brief review of the economic crises that plagued agriculture in the 1980s.

THE ECONOMIC CRUNCH

Farming required a gambler's nerves in the farm economy of the 1980s. The economic shakedown in agriculture in the 1980s was devastating and abrupt, in stark contrast to the prosperity of the 1970s, when export markets expanded dramatically. Between 1981 and 1986, five straight years of accelerating economic decline, the USDA reported that the United States lost 219,500 farms. The foreclosure rate on farm mortgages rose to a shocking 26 percent. In just one year, between July 1984 and June 1985, more than 100,000 farmers had financing cut off. Some farmers managed to sell their farms before they were buried by foreclosure and bankruptcy. In 1985, an additional 44 percent underwent voluntary liquidations for reasons other than normal attrition—presumably, to resolve debts. A combination of surging interest rates, plunging land values, and loss of markets for overabundant farm products wreaked havoc in the farm sector.

In the first half of the decade, debt-to-asset ratios rose dramatically, not because farmers were accumulating debts, but because farm assets in the form of farmland values were dropping. Between 1982 and 1986, land values plummeted 30 percent. This dramatic drop was tied to federal economic policies that changed abruptly with the advent of the Reagan administration. Inflation slowed, interest rates rose, and international markets dried up as the dollar rose relative to other currencies. In 1986, nearly one in four farmers found himself in the doubtful financial position of having debts totaling more than 40 percent of assets. Government payments rose from $8.4 billion to $11.8 billion from 1984 to 1986, which improved farm income enough to disguise some of the stress. Without these payments, more than 50 percent of farms would have had a negative cash flow in 1985. Farming was simply not a self-sustaining business for at least half of all farmers.

Of course, the crisis did not hit all farmers equally. It turned out that "innovative," younger farmers, especially those who began farming in the mid-1970s or later, were the ones in greatest trouble.[1] "Innovative" farmers were the ones who pursued the newest technology and the model of improving efficiency by increasing size. They sought such tax benefits as rapid depreciation of machinery. They tended to have more formal education and to use financial management tools more than farmers with few financial worries. These farmers borrowed on the basis of 1970s land values, as though inflation would last forever. Their debt loads were unsustainable when the 1980s crisis hit. Young farmers were vulnerable because of low equity and high interest payments on newly financed land. When crop prices and land values fell, the land could not produce enough to pay for itself. It became nearly impossible to start up a new farm.

Some farmers who hung on to their land began to look for alternatives to expensive inputs, and low-input farming began to catch on. Financial stress caused farmers to take a second look at the ecology of their farms, for the path to fewer inputs was the path to more ecologically based management.

THE VULNERABLE "MIDDLE"

Mid-size farms, generally what are considered family farms, were also a particularly vulnerable group according to USDA reports. Using sales volume to classify farms into size categories, farms in the two categories with sales from $40,000 to $99,999 and $100,000 to $250,000 had the highest proportion of vulnerable farms from 1984 to 1986. USDA publications generally lump these size groupings into "mid-size commercial farms," or "family farms," sometimes also including farms with sales up to $500,000. This leads to the conclusion that family farms are most financially vulnerable. Conventional wisdom attributes this vulnerability simply to the "awkward" size of farms in this group: large enough to keep the farm family so busy that they must depend primarily or solely on farm income for all living expenses, yet small enough that they don't have enough income-generating enterprises, investments, or dispensable assets to cover losses. To the family farmer, a bad year may mean no household income at all.

This conclusion depends upon how "mid-size family farm" is defined. If one accepts Marty Strange's analysis in *Family Farming* that farms with sales totaling more than $100,000 are more similar to larger farms by a number of criteria and should be grouped with large commercial farms, and that farms with as low as $20,000 in sales should be considered small commercial (family) farms, the numbers say something different. Using these size categories, both large and medium-size farms showed similar vulnerability during the crisis. The USDA classification draws attention away from the brittleness of the heavy debt load carried by large farms. It also helps the figures to better fit the paradigm that "bigger is better." In 1987, the pattern of financial vulnerability among farms changed so that larger farms were more burdened than mid-size farms when considering net farm income or net cash farm income, even when using USDA size categories. This quote from a USDA report on 1987 farm statistics illustrates the propensity of "conventional wisdom" to disguise the facts:

> The overall rates of vulnerability were highest among the largest farms [selling more than $500,000]. Often, though, it is the midsize farms that have the most difficulty handling a serious financial squeeze. Large farms tend to have more assets that can be sold off without jeopardizing the entire operation, and small farms depend more on off-farm income than agricultural earnings.[2]

Nothing more was said about why those large farms were having the most difficulty.

The point here is that the economic crisis of the eighties had a decided effect on the shape, the structure, of agriculture, not just who farms, but how many farmers and how big the farms. And it also says something about what size farms are most resilient, economically speaking. The signs seem to point to a threshold where farms become too large to be resilient and become brittle when economically stressed. These large farms are nearly always the ones that have traveled farthest on the road to industrialization. Chapters 1 and 2 of this book further explore some ecological consequences of industrialization and reasons for brittleness when farms become too large.

As the crisis deepened, and farmers defaulted on loans, lending institutions found themselves holding substantial farmland acreage. Market prices for land were so low that these institutions preferred to hold on to the land in hopes of waiting out better prices and recouping some of their losses. Nationwide, farm acreage owned and operated by insurance companies increased tenfold between 1980 and 1986. Insurance companies are one of the smaller holders of farmland mortgages, but by the end of 1987 they owned 5 million acres (2 million hectares) of farmland. Three million acres (1.2 million hectares) of this lay in the hands of just four companies. Another 1.4 million acres (0.6 million hectares) had been taken over by the Farmers Home Administration (FmHA), and 2 million (0.8 million) more by the Federal Credit System (FCS) by the fall of 1986.

Much of this land was rented and kept in production, but usually not by the original landowner. Juliana King of the USDA's Economic Research Service reported that insurance companies holding farmland employed "in-house management services or outside specialists to get the most return on unwanted assets."[3] Too often this translated to an operator who removed such "extras" as soil conservation measures in order to get the most return on the land. The issue was serious enough to have stimulated a number of concerned farmers in Minnesota to join together in the Land Stewardship Project's Farmland Investor Accountability Program appeal to insurance companies to use conservation plans on their rental land. Thus, financial problems have brought ecological problems into focus. What is to be the future of farmland if economics dictate poor land stewardship?

IN RURAL COMMUNITIES

Shock waves spread beyond the farmers' fences and barns into rural communities. Debt-straddled farmers made poor customers, and so the farm crisis cut into retail sales. Lower sales, in turn, meant fewer jobs in rural communities. As farmers cut back on expenses by reducing input costs, farm machinery dealers and chemical and seed suppliers lost business and scaled back on local employees. Local communities also had to absorb the unpaid debts of bankrupt farmers.

Among 169 North Dakota farmers who liquidated and left the business in the 1980s, 30 percent of their total debt was left unpaid on average, and only 38 percent could pay all their debt after liquidation.[4]

The plight of the farmers foretold the exodus of farm banks that began a few years later. Banks under financial stress passed the pressure on to local businesses as they tightened down on local loans. This downward spiral hit hardest in farm-dependent regions with self-contained banking systems. As of 1983, 42 percent of rural counties were served by such systems,[5] and consequently residents had nowhere to turn when banks tightened down on mortgages and other loans.

Fulda, Minnesota, is an example of a typical hard-hit rural community. Between 1979 and 1984 this town of 1300 people experienced a 55 percent decline in retail sales. As a consequence, five businesses and one bank closed down, eliminating twenty-one jobs.[6] As businesses folded, other residents were forced to look for jobs elsewhere, putting more strain on those who remained to maintain municipal services. The USDA figured that more than half of the nation's small-town counties lost population in 1986. Rural governments' primary source of revenue—local property taxes—declined along with property values. Fulda raised its property tax 8 percent in 1985, and anticipated a further increase the following year, just to make ends meet. Compounding the problem for rural communities was the fact that state and federal revenue sources also dried up during the 1980s.

AND BACK AGAIN

It is a vicious cycle; for the crisis on farms, which spawns crisis in rural towns, returns to the farms to exacerbate the farmers' woes in the form of reduced off-farm employment opportunities and services. USDA figures for 1986 showed that just under half the income in farm operator households came from off-farm earnings. This figure includes the 75 percent of U.S. farms that are very small and obtain 96 percent of household income from off-farm sources. But even if all "small" farms (those that earn less than $40,000 annually from farming) are excluded, and we look at only the mid-size commercial farms (using the USDA definition of farms with sales from $40,000 to $499,000 per year), 25 percent of household income came from off-farm employment. Off-farm income not only influences the stability of farms and the quality of life for the farm family, but tends to recirculate in the local economy. Clearly, farms and surrounding rural communities are economically interdependent.

The economic problems experienced in the 1980s were not merely a temporary aberration. They resulted from long-term trends toward larger, more heavily debt-financed industrialized farms, coupled with chronic overproduction and low prices at the farm gate. They will continue into the future, despite the fact that some indicators improved in the late 1980s. A 1986 General Accounting Office (GAO) publication, *Farm Finance*, predicted that if no further intervention were taken, 23 percent of the nation's farmers would sell some assets to remain in business, and 25 percent more would go out of business before agriculture would stabilize. Since it has become so difficult for young farmers to begin farming, and since the view that "bigger is better" is still strong, the point of stability to which the nation is headed is undoubtedly one of more concentrated ownership of farmland, more industrialized agribusiness, and more economically brittle, heavily indebted farms.

OF ECOLOGY AND ECONOMICS

It has been shown that the economic crisis in agriculture highlights some ways in which economic decisions have ecological consequences. On the positive side, high input costs coupled with marginal incomes in the mid-1980s pushed farmers to seek alternative, low-input, often more ecologically sound ways to achieve soil fertility and pest control. On the negative side, some foreclosed farmland as well as some farmland belonging to economically stressed farmers suffered stewardship neglect.

Chapter 1 presents the ecological side of the farm crisis. Chapter 2 traces the roots of this crisis to farm expansion and the industrialization of agriculture, and beyond—to cultural institutions and value systems that form the current agriculture perspective. It proposes a fundamental shift in perspective that embodies a value system based on sustainability, the idea that farmers can continue to cultivate the land, maintain productivity, and avoid depleting nonrenewable resources into the indefinite future.

Chapters 3 and 4 develop this new ecological perspective further. Chapter 3 compares ecological features of natural ecosystems that provide sustainability with the ecological features of current agricultural systems, and points out basic inconsistencies. Chapter 4 develops the thesis that to achieve sustainable agricultural production, agroecosystems based on models of native ecosystems are needed. The chapter compares process-oriented approaches to a general structural model and, finally, presents a specific proposal for grain production for the prairie region of North America. Chapter 5 presents research showing the feasibility of the proposed new system, a domestic prairie (or perennial, seed-producing polyculture). Chapter 6 strikes a philosophical note as it explores the social aspects of such a fundamental shift in perspective and change in agricultural production systems.

1 _____

Ecological Crises of Modern Agriculture

POLLUTION, DEPLETION, DEGRADATION, erosion, contamination, poisoning—these are terms usually associated with heavy industry and cities, not with our green countryside of fields and scattered farmsteads. But these are all terms that are applicable to the environmental problems caused by conventional U.S. agriculture today. Agriculture has become very like traditional manufacturing industries, with many of the same environmental risks and waste-disposal problems among its side effects. This analogy has a dangerous flaw: because of agriculture's extensiveness and its use of toxins that are broadcast into the environment, its impacts are more wide-ranging than those of most other industries. In fact, they are so widespread that they are generally overlooked. This chapter explores the major ecological problems of agriculture: soil erosion, loss of genetic and biotic diversity, depletion of energy and water resources, chemical contamination of water, workers, and food, and creation of new and more serious pest problems. The breadth and seriousness of these problems present strong evidence that current agricultural practices are ecologically unsustainable.

DAMAGE AND DEPLETION

Despite major awareness stirred up by the Dust Bowl of the 1930s, and more than fifty years of research and government-sponsored soil conservation programs, modern agriculture has failed to produce a system that sustains its own capital—the living soil. Each year more topsoil slips away, down the earth's rivers to the sea. With each harvest, more soil is used up than is rebuilt, as, in effect, the soil is mined from farm fields. Likewise, the other pieces of our agricultural resource base—genetic diversity, fossil energy supplies, and water sources—have been damaged or depleted. In North America this exploitative approach produced the bounty that permitted the rapid colonization and development of an affluent society. Now that humankind has colonized the whole earth, however, people can no longer abandon spent fields and just wander off looking for new soil. As a society we must face the price of our wastefulness and the limits of the planet's ecological foundations.

EROSION

On the hopeful side, soil can be improved, even restored, given enough time. Soil-building processes can be enhanced, and remaining soil can be protected from further erosion. It is not a hopeless problem. But for anything to improve, society must first recognize the seriousness and magnitude of the problem.

On the Farm. Soil erosion increased at an alarming rate in the 1970s. Encouraged by booming export markets and federal farm programs, farmers removed shelterbelts, grassed waterways, and terraces so that they could plant fencerow to fencerow. As millions of acres of erosion-prone land were brought into production, erosion rates rose dramatically. In 1977, the Soil Conservation Service (SCS) initiated the National Resources Inventory (NRI) to document just how much soil was washing and blowing away. They found that, on average, the country was losing 1.8 tons of soil per acre (4 metric tons per hectare) in excess of the "officially tolerable" average annual replacement rate ("T") of 5 tons per acre (11 metric tons per hectare).

In 1982, SCS repeated the NRI, using more comprehensive methods, and came up with even gloomier figures. More highly erodible land had been brought into production during the five years between surveys, and the average loss per acre from croplands was now 8 tons per acre (18 metric tons per hectare), or 3 tons (2.7 metric tons) above T. That meant that 3.4 billion tons (3.1 billion metric tons) of soil were being washed or blown away each year. And that occurred in spite of the fact that acreage farmed with conservation tillage, which reduces erosion, nearly tripled during this period.[1]

Although a small proportion of the cropland accounts for most of the erosion, the majority of the cropland is affected at some level. Twenty percent of U.S. cropland is subject to serious erosion, but fully one third (118 million out of 350 to 400 million acres, 48 million out of 142 to 162 million hectares) is classified as "highly erodible land" (HEL) under the conservation provisions of the 1985 farm bill. The HEL designation means that the land is highly vulnerable to erosion when cropped. While the Conservation Reserve Program (CRP) is removing some HEL cropland from production (25.5 million acres, or 10.3 million hectares, of the pool of 118 million acres, or 48 million hectares, as of February 1988), it is not yet clear how effective this has been at reducing erosion. Regions with the largest acreage of HEL cropland had low enrollment in CRP as of February 1988. For example, only 3.6 million acres out of a pool of 19 million acres (1.5 million hectares out of 7.7 million hectares) HEL in the Corn Belt were enrolled.

Tolerated Losses. What do these sobering figures mean in terms of the depth of soil lost, the productivity sacrificed, and, more importantly, the land's sustainability? The first step is to translate soil loss from tons into inches. It is generally recognized that deeper soils can sustain productivity longer than shallower soils. They can withstand more erosion before they will show much loss in productivity. In Iowa, where soils were deep to start with, it took a loss of 10 inches (25 centimeters) of soil to reduce corn yields by 50 percent. On shallow soils in Nigeria, a loss of only 2 inches (5 centimeters) cut corn yields in half.[2] It takes from 300 to 1000 years for an inch (2.5 centimeters) of soil to be formed naturally. When agriculture causes soil loss to proceed much more rapidly than soil building, the system is literally losing ground; it is an unsustainable system.

If you removed the top inch of soil from an acre of land, it would weigh about 160 tons (145 metric tons). That means that at the national average rate of loss of 8 tons per acre (18 metric tons per hectare) each year, an inch of soil from the average acre of cropland is lost every twenty years (or 1 centimeter every 8 years). That may not sound like a particularly dramatic rate of loss, but when compared to the soil formation rate (300 to 1000 years per inch, or 120 to 400 years per centimeter) it becomes obvious that erosion rates are well beyond sustainable loss. Furthermore, erosion is much higher than average in some regions. At the rate of 100 tons per acre (224 metric tons per hectare) that is lost on some slopes (in some years) in the wheat-growing region of the Palouse in Washington State, it would take only 1.6 years to lose an inch of soil. The average acre in this region loses an inch of soil about every twelve years (14 tons per year, or 12.6 metric tons per year), and this has been going on since the 1920s.[3] Soils were originally so deep and rich in that region that erosion seemed to be inconsequential.

The signs in the Palouse that soils are losing productivity are obvious now. The soil on hilltops is now universally paler than that on the slopes, almost all of which have noticeable rills, many even gullies. The crops are thinner on hilltops too. The apparently limitless rich Palouse soil, originally lying to depths measured in feet, is reaching its limit after only 100 years in production. Still, farmers don't perceive the problem to be as severe as USDA guidelines for the CRP would indicate. In one watershed, 84 percent of the land fell into the USDA HEL land class. Yet farmers only estimated that 24 percent of their land was highly erodible.[4]

The major reason for this discrepancy is probably the fact that the land is still highly productive, and productivity remains the major currency of agriculture. Improved crop varieties and increased fertilizer applications have enabled farmers to disguise soil losses. Yet one must wonder what yields would now be if all that soil were still in place. Could the same high yields now be possible with much less added fertilizer, perhaps? One study of those eastern Washington wheat fields concluded that 90 years of soil erosion has robbed the land of 50 percent of its potential productivity.[5]

Costs on the Farm. Not only does a focus on productivity disguise the fact of soil erosion, but it ignores many other costs of erosion. Failing to recognize these other costs causes society to underestimate the cost of erosion and overemphasize the cost of soil conservation. Erosion clearly increases the direct costs of farming. More fertilizer must be applied to compensate for loss of fertility. Other potential costs of erosion include higher fuel and tractor maintenance costs as soils thin, organic content declines, and the heavier, denser subsoil becomes incorporated in the tillage layer. Erosion increases the risks of farming. Crops may be more vulnerable to drought, disease, and insect damage when growing in eroded soil, which is poorer in organic matter. If all these costs were routinely included in farm budgets, soil conservation practices would be much more popular.

In addition to removing soil, erosion changes the physical structure and biological properties of the remaining soil. Fine particles of organic matter are light, concentrated near the soil surface, and thus the first to go when erosion occurs. This means that the remaining soil has reduced nutrient-holding and water-holding capacity and supports a poorer living soil community. As water-holding capacity declines, soil is less able to absorb rainfall, and more rain runs off the surface. This causes erosion to accelerate. Erosion thus disrupts the natural processes that allow organic matter to accumulate in the soil and to hold soil in place and that perpetuate soil-building processes, while it sets in motion processes that only make the problems worse. The longer erosion is allowed to occur, the faster it will occur, until this waste of "capital" bankrupts a precious resource.

Costs Off the Farm. Significant consequences of erosion occur off the farm as well. Eroded soil is a nuisance, even a hazard, where it is deposited. Reservoirs commonly catch huge volumes of such soil. There the deposition interferes with municipal water supplies and recreational potential. In the United States, 1.4 million acre-feet (0.17 million hectare-meters) of reservoir and lake capacity are lost to sedimentation each year.[6] Dredging to remove sediments is expensive and creates its own problem: what is to be done with all that sludgy material? Municipal water systems bear the brunt of this dilemma, and thus a farm problem becomes an urban problem as well.

Erosion from cropland is the single largest source of nonpoint pollution in the United States, producing about 50 percent of suspended sediments. Sediments interfere with the breeding and feeding of aquatic species, destroying fish, mussels, and benthic insect populations. The problem is compounded when sediments carry nutrients or pesticides with them. Nutrient-enriched waters support algal blooms, reduced oxygen supplies, and rapid aging of lakes.

Flooding also follows erosion, for as the soil's water-holding capacity diminishes, run-off increases, and streams receive more water to carry. At the same time, the sediment loads carried off fields are deposited in riverbeds and wetlands (nature's sponges). As these streams and wetlands become filled with sediments, both the watershed's capacity to hold excess water, and the rivers' capacity to carry it away, diminish. This combination greatly increases the frequency and severity of flooding. Erosion disrupts the natural moderating systems of watersheds and sets into motion processes that are self-accelerating.

Water-caused erosion is not the only problem. Soil particles carried by wind are a significant source of air pollution. Wind erosion accounts for some 33 million to 239 million tons (30 million to 217 million metric tons) of airborne particulates annually. Even the low estimate is larger than the 20-million-ton (18-million-metric-ton) contribution of smokestacks and other point sources.[7] In dry regions, such as the Great Plains, wind erosion is the primary form of soil erosion.

Modern agricultural practices also destroy soil in other ways besides eroding it away. Irrigation can be very destructive to soil, particularly in arid regions. Waterlogging, salinization, and alkalinization have damaged about half of the world's irrigated lands. Once salinization has proceeded to the point that a salty, white crust covers the soil surface, the land becomes unfit for cultivation and can only be reclaimed by extremely expensive procedures. In arid regions, soil and groundwater are typically high in salts, natural drainage is poor, and evaporation is high. With this combination of features, irrigation is a sure formula for land damage. Poor drainage of irrigation water allows water tables to rise and soil to become waterlogged in the crop root zone. Salty groundwater can then seep through unlined irrigation canals to reach the surface, where it evaporates and forms a crust on the soil surface. These problems occur wherever large acreages of cropland are irrigated—from Pakistan to the San Joaquin Valley in California. Worldwide, half of all irrigated lands have been damaged as a consequence of irrigation.

Future Costs. Projections of future consequences of current erosion trends range from the absurdly optimistic view that modern technology is freeing humankind from the need for soil anyway, to the gloomier statistic that at current rates 5 hectares of cropland are destroyed each minute by erosion, worldwide. This amounts to a loss of 3.3 percent of cropland worldwide (3 percent in the United States alone) by the year 2000 (based on 1975 cropland acreage).[8] In *Family Farming,* Marty Strange projects that a continuation of 1980 erosion rates for fifty years would reduce U.S. yield by the equivalent of a loss of 23 million acres (9.3 million hectares). Another estimate for the United States predicts that productivity losses due to erosion over the next fifty years would be equivalent to the loss of 8 percent of the total base cropland.[9] Given a base of 350 million to 400 million acres (142 million to 162 million hectares), this amounts to 27 million to 32 million acres worth of productivity lost (equivalent to 11 million to 13 million hectares). Set against an expected worldwide population increase of more than 50 percent over the same time span, even a 3 percent loss of farmland should be considered intolerable, especially when it is a preventable loss.

The USDA is counting on a combination of retiring highly erodible land into the CRP and disguising productivity losses on the remaining land with new technology. They project that, given another hundred years of 1982 erosion rates, more than half the crop acreage would lose only 2 percent of its productivity and that this could easily be countered with improved crop varieties, additional fertilizer, etc. The trouble is that the CRP program was only set up for a ten-year span, and thinking in terms of a hundred years is still rather short-sighted. To achieve sustainable production, society must learn to think in terms of maintaining the land's productive potential indefinitely.

Despite disagreement on the importance of current erosion rates, most scientists agree that the more land that is brought under cultivation, the more erosion will occur, unless methods change. This is because the best, least erosion-prone lands are already being cultivated. New farmland will be less suited to cultivation and more highly erodible.

LOSS OF GENETIC DIVERSITY

One of the dominant themes of modern agricultural development has been reduction in diversity. This is seen in crop and livestock breeding, where the genetically narrow varieties and breeds that now dominate agriculture have replaced a multitude of locally adapted strains. It is also apparent in cropping systems, as the acreage in continuous monoculture has increased at the expense of acreage in rotation and mixed culture. Even at the global landscape level, conversion of diverse ecosystems to modern-style monocultures reduces the genetic diversity of the earth. Yet diversity is the currency of adaptation. Without it, humans lose the ability to adapt crops and livestock to changing conditions. Likewise, the ecosphere—the living "skin" of the earth—is handicapped in its ability to adapt to change and to remain vital when diversity declines. Declining genetic and species (biotic) diversity threatens the sustainability of agriculture and the resiliency of the ecosphere.

Global Diversity. At its most extreme and perhaps most absurd, global biotic diversity is threatened by the ongoing replacement of tropical rain forests with modern, monoculture-style agriculture. By the late 1980s, tropical forests were being cut at a rate of more than 38,000 square miles (100,000 square kilometers) a year—that is, more than 67,000 acres, or 27,000 hectares, a day.[10] The species extinction rate as these forests come down is conservatively estimated at 4000 to 6000 species a year. That represents a loss of eleven to sixteen species each day that could hold the key to new medicines and new pest resistance, provide a winter refuge for migratory songbirds (birds that provide crop protection for temperate farmers in the summer), or offer a multitude of other benefits to humankind and the planet.

So much is lost, and sadly, so little is gained by such drastic large-scale conversions. Typically, land cleared of tropical rain forests will yield for as little as two or three years and, at most, for only five to seven years of intensive cropping. The reason is that rain-forest soils are very poor in nutrients; it is the vegetation that holds all the nutrients. When tropical forest land is cleared and converted to beef production, productivity rates are pitifully low. Cleared rain forest converted to cattle grazing in Chiapas, Mexico, yields a mere nine pounds of meat per acre per year (about ten kilograms per hectare per year). In sharp contrast, the indigenous Maya use traditional shifting cultivation methods to produce 5270 pounds per acre (5902 kilograms per hectare) of shelled maize, plus 4054 pounds per acre (4540 kilograms per hectare) of roots and vegetables per year for five to seven years, then additional yields of citrus, rubber, cacao, avocado, and papaya for the following five to ten years while the native vegetation is allowed to regrow.[11]

Diversity on the Farm. Ultimately, loss of genetic diversity is a danger to agriculture itself, for the primary source of variation required for crop improvements is locally adapted cultivars or closely related wild species. As farmers abandon the more diverse, locally adapted varieties—called landraces—in favor of the highly selected, genetically narrow, high-yielding new strains, the breeding base of crops is depleted. Global distribution of modern crop varieties has depleted the array of crop-breeding materials. In their chapter in *Agroecology,* Jan Salick and Laura Merrick tell the story of the changes in rice grown in the Philippines. In 1957 the Hanunóo people grew 92 varieties of rice. But in 1981–82, 70 percent of the rice-growing area of the Philippines was planted to a single rice variety, IR36, from the International Rice Institute. While it is true that the new varieties produce substantially higher yields than the average yields of Hanunóo varieties, it is still important to confront the problem of lost diversity and seek solutions that can combine higher yields with higher genetic diversity.

Loss of landraces to new, narrow lines has been occurring worldwide since the green revolution of the 1960s and 1970s. Of greatest concern are regions where crops originated or have been grown for a very long time. For wheat, the critical region is the Mediterranean rim. Greece illustrates the trend in the region: 95 percent of Greek wheat landraces have been lost due to the introduction of high-yielding varieties. Similarly, when high-yielding sorghum hybrids from Texas were introduced into South Africa, nearly all sorghum landraces disappeared.[12] With livestock, the story is similar. Twenty-three breeds of livestock disappeared from Great Britain in this century because of breeders' focus on production-oriented traits.[13]

To understand the drastic reduction in genetic diversity created by this type of replacement, it is necessary to look at just how narrow the genetic backgrounds of the newer varieties are. The origins of the major cultivars of red winter wheat can be traced back to only two original cultivars: Turkey and Marquis. The hundreds of corn hybrids grown in the United States and Canada originated from about twelve inbred lines formed from a few open-pollinated varieties of a single race (out of some 200 known races of corn). And even more extreme, six soybean plants from Asia provided the genetic material for all soybean varieties grown in the United States as of 1980.

The danger of widespread use of a single variety has been demonstrated in dramatic cases of pests breaking through the resistance of a variety and causing widespread crop destruction. Late-blight fungus caused the Irish potato famine, which destroyed half of Ireland's crop in 1846. The entire potato crop of Europe had been developed from only two samples brought back from South America in the 1500s. Both were susceptible to blight. In contrast, many clones are grown in the Andes, where potatoes are native. Though late-blight occurs there also, it does not destroy a major portion of the year's crop, because a mixture of susceptible and unsusceptible clones are grown each year.[14] A similar situation occurred in the United States in 1970 when southern corn leaf blight destroyed 15 percent of the U.S. corn crop because virtually all hybrids were developed from a single source that proved susceptible to the blight fungus.

Ironically, it is the landraces—the locally adapted varieties that the new cultivars have displaced—that are essential to the maintenance of the high-yielding cultivars. Breeders searching for disease and pest resistance, as well as adaptation to climatic extremes and poor soils, turn to landraces first. When that fails, they turn to varieties and related species. Numerous success stories can be found in the literature: powdery mildew resistance in California melons came from a wild melon from India; ripe rot resistance in North American pepper plants was found in a Peruvian species; a Korean strain of cucumber provided high seed yield for U.S. varieties; wild genotypes of rice have helped to boost rice yields. As natural habitats are converted and degraded, there is the danger that sources of wild varieties and related species will also be lost.

Preserving Diversity. Efforts to preserve the genetic resources of crop plants accelerated in the 1970s with establishment of the International Board for Plant Genetic Resources (IBPGR). Centralized gene banks have been established in some 70 nations around the world. The idea is to collect and store seeds of as many varieties and wild relatives of crops as possible, before they are lost. This effort works to some degree, but it is fraught with problems. Seeds die in storage, evaluation of materials for useful traits is a monumental project, retention of vegetatively propagated crops or short-lived seeds requires tissue culture or growing live plants. Grains are fairly well represented, but many fruits, vegetables, and especially tropical plants are neglected, and closely related wild relatives are under-represented. Stored materials are isolated from ongoing processes that maintain genetic diversity. For many crops, the best way to maintain diversity may be to maintain critical aspects of indigenous agricultural practices that produce locally adapted races. But this will require more attention to those practices before they disappear.[15]

Under current breeding and cropping techniques, new varieties, with new disease and pest resistance, are necessary every few years. This requires new genetic sources. There is no reason to expect any abatement in the need for sources of genetic diversity in the future.

In fact, current trends in agriculture are expanding the need. The more widespread the use of uniform varieties in monoculture, the more biological and physical stresses these crops encounter, and the greater the need for new sources of genetic diversity. Expansion of agriculture into marginal lands requires special adaptation of crops to climatic extremes, poor soils, and low water quality. Consumer demands for uniform appearance, flavor, and quality create pressure for new varieties. Changes in farm technology, such as no-till farming, multiple cropping, and biological farming require continued innovation. Pressure on farmers to reduce the use of poisons for pest control will mean even greater reliance on natural sources of resistance. Although all these pressures do not necessarily lead to an increase in crop diversity (uniform characteristics may require less diversity, for instance), they do generally require incorporation of new traits.

The easiest way to preserve the genetic diversity of crops and the biotic diversity of the planet is to keep growing diverse stock in diverse localities and to keep a diversity of healthy ecosystems functioning on the earth. These goals are at odds with the methodology of modern industrialized agriculture. Some combination of crop-breeding methods that produce varieties with greater genetic diversity, cropping systems that incorporate greater biotic diversity, reserves that are designed to maintain diverse stock of crop species, and native ecosystems is needed to halt the loss of genetic and biotic diversity.

ENERGY DEPENDENCY

Agriculture could be described as the process of gathering solar energy into food and fibers with the help of photosynthesis. Yet it would be absurd to call modern agriculture a solar-powered enterprise. A truly solar-powered enterprise is indefinitely sustainable because it is dependent only on contemporary resources. Modern agricultural methods are highly dependent on commercial energy supplements (fossil fuels and electricity, which must be purchased from off-farm suppliers). This supplemental energy has allowed farmers to boost production considerably. It has also given an inflated picture of the earth's capacity to support human life because it is based on an ultimately finite supply of fossil fuel. At this point, in the United States, one can only be confident that the current level of agricultural production is at best as sustainable as fossil fuel supplies, or as the environment's capacity to tolerate the burning of fossil fuels, whichever comes first.

Energy from Farm to Table. Energy dependency is a phenomenon of the twentieth century. During the past century, energy use in U.S. corn production rose eightfold (Figure 1.1). Remarkably, the rising trend did not even slow down in response to the energy crisis of the 1970s, despite dramatic price increases in oil-dependent products. On-farm fuel use did decline by 40 percent, but this was more than

FIGURE 1.1 Energy use in U.S. corn production, 1700–1983.

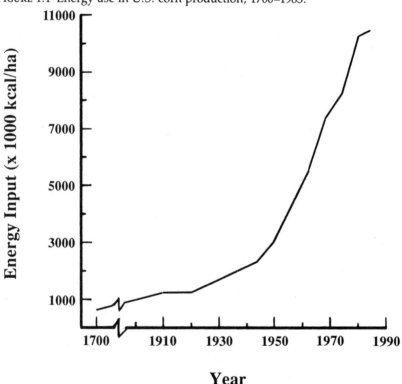

Year

SOURCE: David Pimentel and Wen Dazhong, "Technological Changes in Energy Use in U.S. Agricultural Production," in C. Ronald Carroll, John H. Vandermeer, and Peter M. Rosset, eds., *Agroecology* (New York: McGraw-Hill, 1990), Table 5.2.
NOTE: 1 ha (hectare) = 2.469 acres

offset by the more hidden energy costs represented in increased fertilizer, chemical, and machinery inputs.[16]

Within the context of modern agriculture, nearly all operations are energy-intensive. From traction to shape and move the soil, to fertilizer and pesticides, harvest, threshing and drying—all operations are based on fuels or electricity. Animal muscle power (solar energy) has been replaced with tractor power (fossil fuel). Even seeds are mostly hybrids and require at least the fuel for transport from seed company to farm, since farmers cannot grow their own. Energy is needed to manufacture farm machinery and chemicals, to process and transport feed to animals, and to run other farm equipment.

As long as commercial energy is cheap and abundant, these practices are profitable, despite the fact that more units of commercial energy go into this high-energy agriculture than come out as usable food energy. In their chapter in *Meeting the Expectations of the Land*, Amory Lovins and colleagues calculated that for each unit of food energy produced by U.S. agriculture, two units of energy are used. If one looks only at the food that is actually eaten, the ratio is three to one, and if one includes all the energy costs to ship, process, and prepare food for the table, nearly ten units of energy are spent to obtain each unit of food energy in the United States. The reason that U.S. farmers can afford these huge energy subsidies is that the price of energy in the United States does not reflect all the costs of energy use. Environmental costs of mining and transporting fuels, carbon dioxide emissions, and other air pollutants, for example, are omitted from the equation. Society as a whole presently subsidizes agriculture's energy use by tolerating environmental degradation.

Sharing the World's Supply. Within the context of the current energy-intensive society, agriculture is not a particularly intensive energy user. Energy consumed in farm operations and in equipment and chemical manufacturing amounts to only 3.4 percent of the total energy used annually in developed countries, and 4.8 percent in developing countries.[17] If food processing, distribution, and marketing are also included, the figure is magnified to 17 percent in the United States.[18] When one looks at these percentages, society's massive energy budget masks large agricultural energy inputs, but when looking at the global agricultural energy budget, it becomes clear just how intensive U.S. agriculture is.

The difference between developed and developing countries in total energy use is striking (Table 1.1). Developed countries consume five times the commercial energy for agricultural production as do developing nations. North America alone uses 28 percent of the worldwide agricultural energy budget. Because the developing countries account for the majority of the world's population, the difference in per capita annual energy use is even greater. Developed countries use sixteen times the energy per person-year for agricultural production as do the developing countries. In the United States, the per capita energy used in agriculture is nearly twice that of the other industrialized nations as a whole. Clearly, such inequity cannot be sustained indefinitely. Nor can the rest of the world afford to be as extravagant as the United States is. In their chapter in *Agroecology,* David Pimentel and Wen Dazhong put U.S. energy use into stark perspective when they pointed out that if the world per capita agricultural energy use were equalized in 1984 at the U.S. level, agriculture alone would have used up the world's fossil fuel supply by 1996.

Beyond Dependency. Although agriculture is clearly not the primary cause of energy crises, agricultural energy use threatens the long-term sustainability of our current high yields. On the one hand is the threat of agriculture's comparatively minor contribution to the general environmental destruction associated with fossil fuel use. On the other hand is the threat of vulnerability this dependence lends to ag-

TABLE 1.1

Commercial Energy Use in Agricultural Production

	Total agricultural use (billion Kcal)	Energy per hectare[a] (million Kcal)	Energy per capita (million Kcal)
Worldwide	1817	1.89	14.1
Developed nations	1108	5.92	43.7
North America	511	4.83	79.6
Developing nations	221	0.53	2.6
Centrally planned economies	488	1.41	12.9

Source: B. A. Stout, et al., *Energy for World Agriculture*, FAO Agriculture Series, No. 7 (Rome: Food and Agriculture Organization of the United Nations, 1979), 44–47.
[a]1 hectare = 2.47 acres.

riculture itself. Can the United States sustain anything like current levels of production with current methods when the inevitable end of fossil fuel supplies is reached? How can food production be increased in the bulging rapidly growing hungry regions of the world without major fossil fuel supplements? What would a more equitable redistribution of energy supply around the globe mean for production in the United States and abroad? These questions may be unanswerable in specific, numerical terms. Still, consideration of alternative energy sources and changes in U.S. agricultural practices is a minimal, prudent step.

DEPLETION OF WATER RESOURCES

Modern agriculture is a thirsty enterprise. Although little more than one-sixth of the world's cropland is irrigated, those 680 million acres (270 million hectares) require close to three-quarters of humanity's annual consumption of fresh water.[19] If irrigated agriculture is thirsty, however, it is also productive. That one-sixth of the world's cropland produces one-third of the harvest. Irrigation permits farming in regions that could not otherwise support agriculture and thus increases the earth's capacity to feed the human population. With that population growing at the astounding rate of more than 210,000 per day, (77 million per year) according to the World Commission on Environment and Development (WCED) report *Our Common Future*, it is critical to ask whether this massive use of water for irrigation is a sustainable practice. Unfortunately, in several of the major food-producing regions of the world the answer is no. In major agricultural regions in the United States, the Soviet Union, and China, water use exceeds sustainable levels. The evidence is found in dropping groundwater tables, dependency on fossil water mining, and rivers that are used up before they reach the sea. In these regions, modern irrigation practices exceed the bounds of the natural water recharge cycles and jeopardize both the habitability of whole regions and the future productivity of agriculture.

Mining Water. In the United States, underground water supplies are diminishing as farmers continue to draw on aquifers—natural underground reservoirs—for 40 percent of irrigation water.[20] Aquifers are an attractive source for agriculture because groundwater is usually high quality, more abundant, and a more reliable source than water from rivers, lakes, and above-ground reservoirs. However, in some areas the limits of groundwater supplies are apparent. Groundwater levels are dropping at least a foot a year under 45 percent of U.S. groundwater-irrigated cropland. Texas alone irrigated nearly 4 million acres (1.6 million hectares) from declining groundwater sources in 1982, while water tables dropped up to 4 feet (1.2 meters) in the region (see Table 1.2). These declining levels indicate that water is being pumped out faster than it can recharge by slow percolation through the soil. In the United States, only an estimated 75 percent of annual groundwater usage is replenished. The other 25 percent is simply mined. Texas and California lead the roll of states mining groundwater for irrigation (see Table 1.2).

Water in fossil aquifers—ancient underground reservoirs of water that have been cut off from sources of replenishment—is a nonrenewable resource no matter what the timetable. All use of these fossil sources is mining. Fossil aquifers underlie the Saharan and Kalahari deserts, the Great Artesian Basin in Australia, and the Central Asian basins. In North America, the Ogallala Aquifer is a good example of

TABLE 1.2

States Leading in Irrigation with Mined Groundwater, 1982

State	Area irrigated with declining groundwater in a single year[a] (1000 acres)[b]	Range of annual decline in water tables (feet/year)[c]	Estimated total irrigation use of mined groundwater, 1982 (1000 acre-feet)[d]
Nebraska	1018	0.5–2.0	2200
Kansas	1493	1.0–4.0	3580
California	1592	0.5–3.5	4460
Texas	3814	1.0–4.0	7480

Source: C. Dickason, "Improved Estimates of Groundwater-Mining Acreage," Journal of Soil and Water Conservation 43 (1988): 239–40.

[a]Note that these features reflect acreage actually irrigated in a single year, rather than all acres irrigated in some years (as used in the NRC publication Alternative Agriculture, Table 2–9, p. 113). An even better estimate would average year-to-year variation in acreage irrigated, but these figures were not available.

[b]1000 acres = 405 hectares.

[c]1 foot = 0.305 meters.

[d]1000 acre-feet = 123 hectare-meters.

a fossil aquifer. This vast aquifer underlies parts of seven states: Texas, New Mexico, Colorado, Kansas, Nebraska, Wyoming, and South Dakota. The water was deposited when the Rocky Mountains were young; but the aquifer has long since been cut off from its source of mountain run-off, the climate has become drier, and the aquifer is now overlain with impermeable material in many areas so that even the sparse rainfall of the region cannot reach it. Truly a fossil source, this 3-million-year-old water supplied 20 percent of U.S. irrigated cropland in the early 1980s, according to Sandra Postel of the World-watch Institute.

The Ogallala was first tapped for irrigation in Texas and New Mexico in the 1930s. The devastation of the Dust Bowl had proven that dryland farming was not practical west of the hundredth meridian. Irrigation use increased when pumping technology advanced from windmills to efficient motors and again when the export push of the 1970s and tax policies made expensive irrigation equipment an attractive investment. A sharp increase in irrigated acreage in the major Ogallala states during the 1970s and early 1980s was largely due to increased use of this aquifer (see Table 1.3). One hundred twenty cubic miles (500 cubic kilometers) of water have been withdrawn in the last four decades,[21] and some hydrologists predict that the aquifer will fail to yield in another forty years if early 1980s rates of use continue.

TABLE 1.3

Irrigated Acreage in States Overlying the Ogallala Aquifer, 1969–84

| | Irrigated Acreage, 1000 acres (405 hectares) | | | |
State	1969	1978	1982	1984
Texas	6888	6947	5576	4921
Nebraska	2857	5683	6039	5828
Kansas	1522	2686	2675	2315
New Mexico	823	891	807	674
Oklahoma	524	602	492	440
South Dakota	148	335	376	339

Source: USDA, *Agricultural Statistics* (Washington, D.C.: U.S. Government Printing Office, 1986), Table 540, p. 374.

When the Tap Runs Dry. By the mid-1980s, changing economic trends combined with the exhaustion of the Ogallala to force a decline in irrigated acreage in the region. In Texas, New Mexico, and Kansas, land was taken out of production because of diminishing well yields and rising pumping costs. In a 1983 article in *The Futurist,* Stokes reported that production was threatened on 15 million acres (6 million hectares) in eleven states because of water shortages. Texas alone was expected to lose as many as 3 million acres (1.2 million hectares) by the year 2000. One potentially damaging result of ending irrigation in these regions is a return to dryland farming on a large scale. This would leave vulnerable soils exposed to wind erosion once again. It would also mean lower production.

Whether land is removed from production or placed in dryland production, the next tier of agriculture will be profoundly affected, for Ogallala-irrigated farmland supports 40 percent of the grain-fed cattle produced in the United States.[22] This cattle industry is akin to mining industries in one important way: it is based to a very large extent on a finite resource—fossil water. Careful regional planning might have designated the finite Ogallala as a prudent back-up source to compensate for irregular surface supplies. Instead, overuse of the Ogallala has set up the region for the same sort of boom-bust cycle as has occurred in mining regions everywhere. The ramifications of a bust will surely affect a much wider region as the nation's beef production dwindles or shifts to different regions.

Water Degradation. In addition to simply using up groundwater, irrigation can render water sources impotable. Along coastlines, salt water can intrude from the sea to replace fresh water when the level of water in an aquifer gets too low. Saltwater intrusions have occurred along the California, Georgia, and Florida coastlines.

Surface waters also receive salt loads from irrigation in arid regions. Salt loads increase when water evaporates, concentrating the dissolved mineral salts in the remaining water. Irrigation return flows are thus saltier than the water originally withdrawn and add to the saltiness of rivers. Simply removing high-quality water from a river leaves less water to dilute the natural salt concentration that results from evaporation in the river itself. In major river basins used for irrigation in the western United States, salinity increases dramatically from headwaters to mouth because of irrigation depletion, canal seepage, and return flows. The salinity of the Rio Grande in Texas rises from 870 to 4000 milligrams of salt per liter in just one 75-mile stretch (120 kilometers) of its course, according to Mohamed T. El-Ashry and colleagues in "Salinity Pollution from Irrigated Agriculture." Besides the infamous salty Colorado, other affected river basins include the Yakima in Washington, the Snake in Idaho, the Arkansas in Colorado, and the Pecos in New Mexico.

Minerals and salts can also be picked up directly from soil when water seeps through irrigated soil and eventually returns to groundwater, streams, or subsurface drainage systems. A dramatic case occurred in the southern San Joaquin Valley of California in the 1980s. The Tulare Basin has no natural outlet, and its soils are naturally rather high in salts. To keep soils productive in this region, subsurface drainage was necessary. An irrigation drainage system was begun in 1968 but never completed because funding ran out. Instead of draining into the Sacramento–San Joaquin River Delta system, as originally planned, the San Luis drain ended in the Kesterson Reservoir (subsequently designated a wildlife refuge), and the reservoir began to receive drainage water in the late 1970s. This proved to be a deadly arrangement for the wildlife of Kesterson, for the drainage water had picked up the mineral selenium as it passed through irrigated soils in a 42,000-acre (17,000-hectare) portion of the Westlands Water District. The selenium reached a toxic level as reservoir waters evaporated.

In 1981, Kesterson's new wildlife manager wondered why there were no muskrats, crayfish, or turtles in the reservoir. By 1982, waterfowl began showing grotesque birth deformities and most fish species died out. This tragic incident dramatically brought the hazards of subsurface irrigation-drainage waters to the public's attention and prompted a great deal of debate about how to manage salinity and irrigation in California. Kesterson has not received any subsurface drainage water since June 1986, but the question of how to practice irrigated farming safely in this and other arid regions is far from resolved.

Off-Farm Versus On-Farm Use. Irrigation withdrawals from rivers and streams are often extreme enough to damage riparian habitats. The figures are dramatic. In the eighteen western states most dependent on irrigation, 70 percent of the instream flow is often depleted. By the time the Colorado River reaches its lower reaches, more than 80 percent of the instream flow has been withdrawn. In the Upper Colorado River watershed, 70 percent of withdrawal is for irrigation.[23] Rivers of fourteen western and southwestern states are approaching critical levels for survival of fish and wildlife habitats. To protect these natural habitats would require an estimated 80 percent reduction in irrigation. Since most of our irrigation is less than 50 percent efficient, this is perhaps not as outrageous as it may initially sound.

Irrigation efficiency and reuse rates are becoming major issues in areas where agricultural water use is in direct competition with urban water use. Urban use not only accounts for a much smaller percentage than agricultural use, but some 90 percent of water used by homes and industry can be recycled for reuse. In agricultural use, not only is a great deal of water lost to inefficiency, but very little of it is returned to the source for reuse. Estimates for the efficiency of irrigation water use range from 37 percent to 57 percent on a global basis. Return flow for reuse estimates range from none to 50 percent of the initial withdrawal. Agriculture offers the most potential for conservation in areas where water sources are critically overburdened and urban and agricultural uses conflict. In Tucson, Arizona, where agricultural use comprises 90 percent of all water use (groundwater in this case), a 10 percent increase in irrigation efficiency, resulting in 10 percent less water used, would double the water available for urban use.[24] Likewise, where more water is needed for agriculture, increasing efficiency would help. In the Indus region of Pakistan, it would take only a 10 percent improvement in efficiency to provide enough irrigation water for an additional 2 million acres of farmland.[25]

A great deal of the recent dramatic gain in production of grains— for example, corn and sorghum on the western Great Plains—has come about because of an increase in supplemental irrigation.[26] High yield was achieved by essentially altering environments so that crops could be grown in inappropriate climates. This strategy has certainly been effective in raising agriculture's production statistics, but inasmuch as it relies on mining groundwater or salinizing soil and water, it amounts to padding the statistics and clearly has a limited future. It is critical to ask how much longer society will be willing and able to spend 70 percent of the earth's fresh water on growing crops with little respect to their natural climatological limits. A sustainable agricultural system for the future will surely include irrigation in some regions, but it must strive to minimize wasteful irrigation practices and work within the bounds of the earth's natural water-recharge cycles.

CHEMICAL CONTAMINATION

Thirteen pounds of pesticides, 140 pounds of nitrogen, 42 pounds of phosphate—these were the average chemical inputs per irrigable acre in California in 1980, according to Michael McKenry in his chapter in *Pesticides in Soil and Groundwater.*[27] Total U.S. pesticide use on croplands was about 800 million pounds (360 million kilograms) in 1978. Total fertilizer use, including substantial off-farm use, peaked at about 77 million tons (70 million metric tons) in 1981. Along with this massive broadcast of chemicals onto farm fields has come a complex array of environmental problems. These can be broadly categorized into external and internal problems. External problems are those that primarily affect the world beyond the agroecosystem: pollution of ground and surface water and human exposure to pesticides through foods and farm work. Internal problems are those that primarily affect agroecosystems themselves: pest and disease outbreaks caused by the use of chemical pesticides.

CONTAMINATED WATER SUPPLIES

The 1980s will go down in environmental annals as the era of toxic wastes. It was the decade when the major federal push to clean up toxic-waste sites was initiated, as well as the decade when it became clear that even agriculture had a toxic-waste problem. People had been aware for some years that nitrate could get into groundwater and cause problems. But in the 1970s scientists were still debating the extent to which nitrogen fertilizer could be implicated as a major source of this pollution. By the mid-1980s numerous careful studies had eliminated the debate: it was not just the heavily concentrated sources, such as feedlot manure and human-sewage disposal sites, that were polluting; the miracle drug of farming, inorganic nitrogenous fertilizer, was a major culprit too.

Nitrate in Groundwater. Worldwide, numerous studies have shown that nitrate concentrations in groundwater are strongly correlated with overlying land use.[28] In the Big Spring Basin, Iowa, George Hallberg undertook a long-term study of the entire groundwater basin. He documented a threefold increase in nitrate concentration in the groundwater over the twenty-five-year period from 1958 to 1983. The only major change in nitrogen sources in the basin during this period was a threefold increase in the amount of nitrogen fertilizer applied to the farmland. As fertilizer application increased, so did ground-water contamination.

As crops are overloaded with nitrogen fertilizer, the ecosystem's capacity to put the nitrogen to use is soon exceeded. The soil organisms process the excess into soluble nitrate, which gravity carries down below the root zone in a process called leaching. Nitrate is usually quite stable below the root zone because biological activity is minimal there. Although it is possible for denitrification to occur in groundwater, the capacity is probably not often high. Once the soil's natural safety filter is overloaded, serious pollution problems begin to develop.

Too Much Fertilizer. A great deal of nitrate does get through to farmland subsoil, and the primary reason is that farmers have been applying more fertilizer than their crops can use. In fact, on the average, about 50 percent of the fertilizer applied in the United States is not used by crops. The degree of groundwater contamination is often closely related to the excess nitrogen fertilizer applied—that is, the amount beyond the nitrogen harvested in crops. The notion that "more is better" has gotten out of hand here. Corn plants take up only about 50 percent of the fertilizer typically applied, and the harvested grain accounts for only about 35 percent.[29] Part of the problem is in timing the application. If fertilizer application is timed to coincide with a crop's active growing and flowering periods, crops will use fertilizer more quickly and completely. This is not the most convenient time for the farmer, however.

Carefully planned crop rotations can minimize nitrate leaching. A rotation should include deep-rooted crops following shallow-rooted crops to recover nitrate from deeper soil zones. A biculture of the two types could make more complete use of applied nitrogen. In India, growing deep-rooted crops during the rainy season and shallow-rooted crops at other times of the year reduced nitrate leaching.[30] In Great Britain and other areas with similar climate, a fall cover crop and omitting a fall fertilizer application are the best preventions.[31]

The nitrate issue is an emotional one—partly because nitrate is toxic to infants and perhaps because it goes against our sensibilities that something as helpful as fertilizer can be harmful too. The maximum safe concentration in drinking water is set at ten milligrams per liter (or ten parts per million). At concentrations of forty-five parts per million, it can cause fatal methemoglobinemia sickness, or "blue-baby syndrome," in infants receiving formula made from contaminated water. In 1986, an infant's death in South Dakota was attributed to fertilizer contamination of her family's well water.[32] Interestingly, this incident received virtually no press coverage outside the Midwest.

A number of surveys have been published with estimates of the extent of groundwater pollution by nitrate. The percentages vary greatly (see Table 1.4), from 20 percent to 70 percent of sampled wells, but even the low-end figures are alarming. In *Ground Water Quality and Agricultural Practices,* Deborah Fairchild reports that thirty-one states in the United States have regional or local problems of fertilizer-contaminated groundwater. From European nations to Israel and India, the story is similar. Wherever modern agricultural methods based on intensive inorganic nitrogenous fertilizer or inten-

TABLE 1.4

**Extent of Groundwater Contamination with Nitrate,
from Selected Studies**

Location of survey	Percentage of wells contaminated	Year of survey
Washington County, IL	73	1972
Merrick County, NE	70	1976–77
14 counties, IA	40	1978–81
47 counties, IA	>20	1978–81
Alliston, Ontario	40[a]	1981
North-central IA	9, 20, 37[b]	1982–83
12 counties, NE	22	1984
Nationwide, U.S.A.	20	1985[c]
Statewide, KS	28	1986

Sources: Singh and Sekhon, "Nitrate Pollution of Groundwater," 209; Hallberg, "From Hoes to Herbicides," 359; Fleming, "Agricultural Chemicals in Ground Water," 124; Hill, "Nitrate Distribution," 699; Baker, Kanwar, and Austin, "Impact of Agricultural Drainage Wells," 518; Chen and Druliner, "Agricultural Chemical Contamination," 105; Knudsen, "Experts Say Chemicals Peril Water," 8.

[a] Percentage of study area underlain by contaminated groundwater.
[b] Percentages from each of three study areas.
[c] Publishing date; study was a compilation of data from many other studies.

sive animal husbandry have been employed, nitrate pollution of groundwater has followed.

A Nitrate Legacy. As the chemistry involved in nitrate leaching has become better understood, it has become clear that this is a problem that won't go away overnight. Agricultural soils are loaded with stored excess nitrogen, which can slowly leach into deep aquifers, even for years after fertilizer application ceases. The average time needed for nitrate to percolate from the soil surface to a depth of 49 feet (15 meters) was ten years in one California study.[33] According to computer simulations, it could take another forty years to see the full extent of deep groundwater contamination from current agricultural practices.[34]

Groundwater decontamination is usually technically and economically impractical. It will have to take place naturally by recharge with uncontaminated water, and this happens just as slowly as contamination. The checks and balances in natural ecosystems work both ways. For example, a naturally slow transit time to groundwater permits enough time for recovery of nitrate that leached during growth of a shallow-rooted crop by growing a deep-rooted crop the next season. This "check"—the slow transit time—can work to protect groundwater. But when farmers consistently overload the system year after year with excess nitrogen, the same check works against safe groundwater, by increasing the time for decontamination and making it harder to regain a stable, healthy ecosystem.

Pesticides in Groundwater. In the latter half of the 1980s, concern about nitrate was overshadowed by another newly discovered threat to groundwater supplies—pesticides. Organic pesticides, the modern replacement for arsenic-based chemicals, grew out of the industries that worked on chemical warfare; the most notorious of these pesticides was DDT, a typical chlorinated hydrocarbon. This class of pesticides was usually long-lived, of low toxicity, and bioconcentrating. The latter characteristic refers to the property of staying in biological tissues, where it becomes more concentrated at each trophic (feeding) level in the food chain. By the early 1960s, the consequences of this property began to emerge and were most dramatically reported in Rachel Carson's *Silent Spring.* Carson shocked us with figures such as fish found with 100,000-fold DDT bioconcentration. The effect on top predators, such as bald eagles, was devastating. The DDT caused eggshells to be thinner, which dropped reproductive potential to near zero in many regions. Slowly, the effects of long-term exposure to such chemicals were revealed to be mutagenesis and carcinogenesis. In 1969, DDT was finally banned, and many other similarly long-lived pesticides were banned in the early 1970s. The concerns about bioconcentration and the long life of chlorinated hydrocarbons are clearly well founded. Residues of these toxins are still present all over the planet, in all populations, some twenty years later.

Chlorinated hydrocarbons have been largely replaced with several types of short-lived chemicals, including organophosphates (such as malathion) and carbamates (such as aldicarb). Although these are not prone to bioconcentration, they are much more acutely toxic, so that even minute quantities can be harmful. For a time, these newer chemicals seemed to mark the end of the problem of pesticides getting into the water and food chain.

From Farm Fields to Wells. It was not until 1979 that routine agricultural use of the new generation of pesticides was linked to groundwater contamination. That was the year that aldicarb—marketed as Temik—was found to have leached through the sandy-soil potato fields of Long Island, New York, into the groundwater. Aldicarb affects the nervous system, causing temporary cholinesterase depression when exposure is sublethal, and is acutely toxic to humans. It is used to kill insects and nematodes and is applied to the soil. More than a thousand wells were contaminated. Union Carbide, manufacturer of Temik, quickly removed the product from sale on Long Island and installed filters to clean well water.[35] At about the same time, DBCP, another soil-applied nematicide, began to show up in wells in the San Joaquin Valley of California, even though it had been banned two years earlier because of severe health risks (sterility and cancer) to manufacturing workers.

As the 1980s unfolded, it seemed that the more that researchers looked, the more they found. The nematicide EDB turned up in groundwater in Florida, Georgia, and Hawaii. Wisconsin's central sandy areas also exhibited aldicarb contamination. All these areas had one feature in common—sandy soils overlying shallow water tables—and thus it was not surprising that groundwater contamination occurred there first. Part of the reason for finding more and more pesticides in the water was that scientists were looking more, part was that more sensitive tests were being developed, and part was a matter of increased pesticide use. Herbicide use in 1982 was nearly 2.5 times the 1971 level. Overall agricultural pesticide use more than doubled from the 1960s to the 1980s. But another factor in the increased detection of pesticides may have been the passage of time. In the years following the first detections of pesticides in shallow groundwater, pesticides began to show up in deeper wells too. The time lag was probably related to the fact that it takes longer to percolate to deeper aquifers.

Researchers are just now discovering that apparently short-lived, unstable compounds can become extremely tenacious once below the biologically active soil zone, or even when trapped in topsoil micropores, where the usual biological degradation does not occur. The case of EDB, ethylene dibromide, is a good example. Once they were thought to leave no residues, but more sensitive tests proved that EDB could remain in soil for many years. In Connecticut, this nematicide was discovered in the topsoil of a field where it had last been used nineteen years previously.[36] Persistent residues offer a continuous source of chemicals for slow leaching into groundwater. Scientists are finding that the capacity of structurally and biologically complex soils to detoxify these chemicals is not simply predictable from laboratory tests and varies considerably with soil type, farming practices, organic matter, and weather. Uniform processing capacity cannot be assumed, as it can in a standardized industrial-detoxification scheme. Just as with nitrate, complete cessation of pesticide use would not immediately halt the increasing presence of pesticides in groundwater.

An EPA survey in the late 1980s reported agricultural pesticide contamination of groundwater in at least twenty-six states, involving at least forty-six different pesticides. In *Balancing the Needs of Water Use*, James Moore estimated that tainted well water was the drinking source for 63 percent of rural Americans. Although pesticide concentrations tended to be quite low, these findings should not be dismissed as of little or no importance, because the health consequences of long-term, low-dose exposure are still unknown, particularly for children.

From Farm Fields to Surface Waters. Pesticides contaminate surface waters also. Communities that rely on surface water sources—such as reservoirs, lakes, and streams—can sustain intense exposure at times, particularly when heavy rains follow field applications. Samples taken after spring rains in Iowa turned up pesticides in the water supplies of thirty out of thirty-three communities relying on surface sources. In a heavily irrigated area in central Asia, 74 percent of water samples contained organophosphates and phenoxy herbicides.[37] Although concentrations are usually low, the impact is unknown. Only one of the thirty-seven compounds included in the Iowa study (the herbicide 2, 4-D) had been assigned a safety standard under the Safe Drinking Water Act. At least two chemicals, atrazine and metachlor, clearly violated lifetime health advisory levels set by EPA when at their peak concentrations after rains, but it is not clear how dangerous this may be. Atrazine is highly soluble and long-lived and showed up most frequently in water samples in a nationwide river survey conducted from 1975 to 1980. It has been implicated as a carcinogen, but the EPA recently refused to study it further, citing lack of convincing evidence of carcinogenicity. In the midwestern United States, triazine herbicides (such as atrazine) have become commonplace in stream water. U.S. Geological Survey figures released in 1990 showed triazines in 50 percent of streams prior to spring field applications. After spring applications, 90 percent were contaminated, and 34 percent of the samples exceeded EPA drinking water standards for the herbicide alachlor. The impact of intermittent high concentrations, or long-term low concentrations, of pesticides on humans or on aquatic ecosystems is simply unknown.

Health Effects. While pesticides continue to seep toward groundwater supplies, human exposure rates continue to climb. While the EPA debates "safe" levels of contamination, epidemiological studies in the Corn Belt show elevated cancer rates among farmers (see Table 1.5). Throughout rural America, agricultural chemical use is associated with elevated rates of certain cancers.[38] In Iowa, higher cancer rates were found to be associated with a high rate of herbicide used at the county level.[39] In Wisconsin, higher cancer rates were found in counties with intensive insecticide use.[40] In California, the contamination of 10,000 wells with the nematicide DBCP (now banned) has been linked to elevated levels of cancer of the esophagus and stomach and with lymphoid leukemia.[41]

Some states are not waiting for the EPA to act. After the Iowa Geological Survey discovered pesticides in 39 percent of 500 wells sampled and elevated nitrate in 25 percent, Iowa passed its contro-

TABLE 1.5

Cancer and Farming: Selected Studies Showing Elevated Cancer Mortality Among Farmers and Aspects of Farming That Correlate with Cancer Rate

Location of study	Date published	Years of death	Type of cancer	Best predictor
Nebraska	1979	1957–74	Leukemia	Heavy corn production in county
Wisconsin	1982	1968–76	Non-Hodgkin's lymphoma	High acreage in small grains and insecticide use in county
Iowa	1983	1964–78	Multiple myeloma	High herbicide use in county[a]
			Non-Hodgkin's lymphoma	High production of egg-laying chickens in county[b]
			Stomach	High milk, cattle, corn production in county
Wisconsin	1984	1968–76	Multiple myeloma	High acreage treated with insecticides in county

Sources: Blair and Thomas, "Leukemia among Nebraska Farmers," 264; Cantor, "Farming and Mortality from Non-Hodgkin's Lymphoma," 239; Burmeister et al., "Cancer Mortality and Farm Practices," 72; Cantor and Blair, "Farming and Mortality from Multiple Myeloma," 251.
[a]For those born after 1900.
[b]For those born before 1900.

versial Groundwater Protection Act of 1986. The act takes a new approach by taxing potential sources of groundwater contamination. It requires wholesalers to purchase stickers for household products that produce hazardous wastes and taxes chemical storage tanks and garbage dumps. The revenues generated are used to pay for regular monitoring of wells and research into safer methods of farming.[42] In 1988, funds from the act helped to support a project that reduced fertilizer and pesticide applications on fifty farms in Butler County by providing soil tests and weekly pest monitoring. In Nebraska, the state legislature gave local soil conservation districts the power to restrict farm chemical use in groundwater recharge areas.

The future promises increasing public objection to the agricultural use of chemicals as the environmental consequences become more obvious and well documented. Agriculture has exceeded biological systems' capacity to process the foreign chemicals it employs, and the result is instability of the larger support system, the water-purifying system of the soil. In the case of agricultural chemicals, the instability is manifest as unacceptable health hazards to people not necessarily involved in agriculture, whereas with erosion it is manifest as destruction of agriculture's living capital—the soil. Agriculture cannot continue indefinitely to use these chemicals as it now does. Political pressures will demand a change in this unsustainable practice.

FARM WORKERS AND PESTICIDES

Whereas the discovery of low levels of quantities of pesticides in groundwater supplies has been receiving wide press coverage, much less has been said about the hazards for people such as farmers, field laborers, greenhouse and nursery employees, and pesticide applicators, who are routinely exposed to pesticides in their daily work. Agriculture has one of the worst records for injury and death of all U.S. industries, ranking among the top three, along with construction and mining.[43] In California, agricultural workers have twice the average occupational disease rate of all other industries combined. Yet we do not know what proportion of disease and injury is caused by pesticide exposure.

Assessing the Risk. It is difficult to assess the extent of pesticide poisoning because symptoms of low-dose exposure can easily be confused with other diseases, and because legal incentives for reporting do not exist in most states. Subacute exposure to organophosphates and carbamates produces flulike symptoms: headache, nausea, drowsiness, confusion, and depression, all easily misdiagnosed if doctors are not trained to recognize them or to ask questions about occupational exposure. But the effects can be more serious than they sound. The symptoms are caused by depression of cholinesterase enzyme activity, which interferes with nervous system functioning. More acute levels of exposure cause convulsions, paralysis, and death. In the case of organophosphates, each successive exposure further depresses the cholinesterase activity and, in effect, lowers the acute toxicity exposure level for the worker. A few researchers have set out to determine if agricultural workers are being dangerously exposed to pesticides by measuring their cholinesterase activity levels. Agricultural workers in California, New Jersey, Canada, and Nebraska showed signs of subacute exposure to organophosphates and carbamates. The degree of exposure among field workers was found to be greater during periods of pesticide applications and higher than that of other kinds of workers.[44] Long-term health problems of acute poisoning include probable permanent neurological damage, birth defects, reproductive effects, and genetic damage.

Physicians are not required to report pesticide illness in most states, and the exclusion of farm workers from Workers' Compensation in many states means that doctors have no incentive to report. California has the strictest reporting law in the United States: doctors are required to report pesticide-related illness within twenty-four hours or face a $250 fine. Still, one California State Department of Health study described by Ephraim Kahn in "Pesticide Related Illness in California Farm Workers" estimated that as few as 1 percent to 2 percent of the cases are actually reported.

Nevertheless, in "The Health Effects of Agricultural Production: I. The Health of Agricultural Workers," Molly Coye provides statistics that indicate that the problem is significant. Agricultural workers reported fifteen times more doctor visits for symptoms that were potentially pesticide related than a control group of the same ethnic and socioeconomic background. In 1982, 235 cases of pesticide-related illness were reported in California. If the estimated reporting rate of 1 percent is correct, this represents 23,500 cases among about 300,000 farm workers. That is about an 8 percent illness rate. Applied nationwide to seasonal farm workers, assuming this is the most vulnerable group, this means a conservative estimate of some 156,000 cases of pesticide-related illness per year. The cost in terms of worker productivity and medical expenses is rarely acknowledged. David Pimentel and colleagues, in a paper titled "Pesticides: Environmental and Social Costs," estimated the annual economic costs of human pesticide poisonings in the United States at more than $180 million. This estimate was based on a much lower rate of poisoning (45,000 cases per year) than that derived here, and therefore costs may actually be many times greater, even without considering all the nonmonetary costs.

The workers most heavily exposed to pesticides are those who harvest or tend perishable fruits and vegetables. These are usually seasonal migrant workers with low incomes, inadequate diets, and such stressful working conditions as lack of fresh water. Sensitivity to pesticides is increased by malnutrition, dehydration, and stress. Those who actually apply the chemicals are also at risk, but here regulations may have more effect, since applicators must be licensed and thus, to some extent, trained.

Other populations at risk of direct pesticide exposure are farmers, already noted above to have elevated risks of certain cancers; farm neighbors, who are exposed to drift from aerial, or even field-applied, sprays; and even suburbanites who use lawn chemicals. The exposure of farm neighbors is becoming an increasing problem as urban sprawl continues to encroach on the countryside and housing developments are built adjacent to farm fields. Exposure of suburbanites, of course, gets more public attention than do the problems of farm workers.

Paying the Price for Perfect Produce. Perhaps the ultimate story of exploitation of farm workers for the sake of our cosmetic quality standards comes from Mexico. On the west coast of Mexico, in the state of Sinaloa lies the Culiacán Valley, which, with the help of extensive irrigation and intensive pesticide use, supplies one-third to two-thirds of the fresh tomatoes and other vegetables eaten in the United States in winter months. Culiacán has a bad reputation among the 140,000 to 200,000 migrant workers who cultivate, pick bugs, spray pesticides, and harvest the crops. As one man put it, it is *"una tierra de enfermedades"*—a land of sickness.[45]

Angus Wright observed farm practices in Culiacán in 1983–84 and reported that it is standard procedure there for pesticide applicators to use no protective gear. Most wore sandals, did not wear gloves or masks, and often carried leaky tanks of chemicals on their backs. Even though Mexican growers now use mostly U.S.-approved short-lived pesticides, in Culiacán pesticides are applied so frequently— every two to four days from mid-January to May—that workers experience nearly continuous exposure.

Estimates from emergency care clinics in Culiacán show an average of ten to twelve cases of pesticide poisoning per month, or up to one per day during peak spraying months. However, pesticide poisoning is officially recorded as "other illness" (a category with 1000 cases a month in the state of Sinaloa); so the actual numbers are unknown, but are certainly much higher than officials like to admit.

Since those most severely affected by direct pesticide exposure are a relatively powerless minority, this deadly effect of agricultural pesticide use poses no immediate economic threat to our agricultural system per se. However, to those whose awareness has been raised, it is a grisly price to pay for perfect produce and fresh vegetables year-round. And it is particularly ironic for agriculture, which ought to be in the business of providing sustenance and health to the whole population, to be engaged in poisoning field workers. This irony extends to another aspect of pesticide hazards—the issue of food safety.

PESTICIDES ON FOODS

Though perhaps the least of the hazards of pesticides in terms of the number of lives endangered in the United States, pesticide contamination of food loomed large in the public consciousness in the 1980s. In 1984, Chris Lecos reported in "Pesticides and Food: Public Worry No. 1" that pesticides on foods, particularly imported foods, were consumers' number-one concern about food safety.

Public Power in the Marketplace. One of the flaws of pesticides and other agrichemicals is that they are often hard to get rid of once applied to food crops. In fact, this may prove to be the fatal flaw of these superchemicals. Little can raise public emotion like the knowledge that our food is tainted with toxic chemicals, and the economic pressure brought to bear by public reaction is powerful indeed. The sensational 1989 reporting of the dangers of the growth regulator Alar (daminozide) on apples had apple growers panicking as sales dropped dramatically, despite the fact that most growers claimed to have stopped using the product three years prior. When the suspected carcinogen EDB, a fungicide used on stored grains and milling equipment, showed up in Pillsbury cake mixes in 1983, public reaction was swift. Products were yanked off grocery shelves, and suddenly the EPA was spurred to speed up actions to ban the chemical for use on stored grains. The agency had been sitting on information about the dangers of the chemical's residues for seven years without taking action (largely because of pressure from manufacturers and users) until the public got wind of it.[46]

EDB had been used since 1948 as a soil fumigant to kill nematodes, on citrus crops, as a post-harvest insect control for stored grain, and as a fumigant for milling equipment. Until more sensitive tests were developed in the 1970s, it was thought to leave no residues. When the pesticide began showing up in the groundwater in Florida, Georgia, California, and Hawaii, the EPA took action, issuing an emergency ban of EDB use as a soil fumigant in October 1983. Following the cake mix scare, the agency also banned its use on grains and milling equipment, and finally on citrus crops, in early 1984. The final tally of costs of these actions includes Pillsbury's insurance claim of $50 million in damages due to business losses caused by publicity about the company's EDB use,[47] somewhere between $2.5 million and $8 million in costs for EPA disposal of the remaining stocks of the chemical, untold health costs from long-term exposure, as well as incalculable costs of polluted groundwater for years to come. The power of public response to agricultural practices, as expressed in the marketplace, should not be underestimated.

Although our cake mixes and oranges may now be protected from direct EDB exposure, other environmental effects will persist. The chemical has proven to be extremely long-lived in soils. One researcher estimated its half-life in groundwater as 300 to 500 days, so it will not be eliminated soon from our environment.[48]

Extent of Exposure. The Food and Drug Administration (FDA) admits that pesticide residues do show up frequently in our food, but by the time the foods are prepared and actually reach consumers' tables the residues are "consistently" well below permissible levels of intake.[49] It should be noted, however, that this conclusion is based on surveys using small sample sizes.

The long-lived pesticides such as DDT, other chlorinated hydrocarbons, and parathion are most likely to leave persistent residues. Even though most long-lived pesticides are banned in the United States and other wealthy countries, they are still used elsewhere, or are still persistent in the environment where previously used, and consequently show up frequently on foods. Other pesticides, such as most organophosphates, tend to break down so rapidly that they are unlikely to show up on food unless applied to crops very close to harvest time. It takes a much smaller dose of these chemicals to be acutely toxic, however. The chemical DDE, a breakdown product of DDT, was still found in about 40 percent of the food sampled from U.S. grocery shelves in 1983. The presence of this chemical, along with other chlorinated hydrocarbons, in food has declined steadily since the 1970s. During the same period, however, amounts of organophosphates have increased, corresponding to their increased usage. For example, malathion was found in about 20 percent of the food sampled in 1983.[50]

Approved pesticides are supposed to leave little residue on crops when used according to directions. National and international standards of acceptable residue levels are based on approved usage and maximum acceptable daily intake (MADI) levels set by the Food and Agricultural Organization (FAO) and World Health Organization (WHO), so that theoretically foods should be safe. However, it is not practical to monitor all crops and shipments; residues are not always detected even when tests are done; pesticides are not always used according to directions; and therefore dangerous residues do make their way to the food people eat.

Exporting Pesticides and Importing Residues. Although the FDA maintains an active monitoring program that makes it advantageous for growers exporting to the United States to comply with U.S. standards, only a small percentage of shipments can actually be tested. According to FDA figures, some 2000 food shipments crossing the Mexican border are tested annually. This is indeed a small percentage of the Mexican import traffic, which exceeds 900 truckloads per day during the peak of the January to April shipping season. Although the tests theoretically can be performed in two days, a 1986 GAO investigation found that the average time was eighteen days. Rather than holding perishable produce at border checkpoints while awaiting results, officials routinely allow shipments to pass. Positive tests, indicating dangerous pesticide residues, are found in about 6 percent of the shipments. Only a little more than half (55 percent) of the illegal shipments were prevented from reaching consumers, and in only a few cases were damages collected from offending shippers.

In less developed countries, especially in tropical regions, pesticide use is not regulated as carefully as in the developed countries. Cases of misuse of pesticides are common because directions for proper use generally are not available. Labeling is poor and often not in the native language because many pesticides are produced in the United States and other developed countries and exported to developing nations. Extension advisory services are usually unavailable, and proper usage outside of temperate zones may be unknown, even by manufacturers. A number of pesticides banned in Western developed countries are still used in less developed parts of the world because they reportedly degrade faster and are therefore safer in tropical climates. The result is that exposure to pesticides in foods is much more extensive and intensive in less developed countries than in the United States.

Too often this means that the people in less developed nations are consuming dangerous amounts of pesticides. Y. P. Gupta reported on pesticide misuse in India. In one market in India, a seven-year study showed organochlorine pesticide residues above tolerance limits in more than 30 percent of the food. The average Indian has between 12 and 31 parts per million DDT in his or her body tissues compared to 3.3 parts per *billion* for people in the United States. Twenty percent of Indian farm workers and 75 percent of Indian pesticide manufacturing workers show signs of severe poisoning. Unfortunately, it is all too easy to compile a gloomy list from around the globe.[51]

The stories go on and on, and consistently it is the poor who are most severely affected. Foods exported from less developed countries are usually monitored so that they will pass the inspections that the richer importing nations impose. Locally marketed foods in the less developed countries usually are not monitored; yet these are the products most likely to have residues of short-lived, highly toxic organophosphates, because the time between harvesting and marketing may be shorter than the time it takes for the pesticides to degrade. Thus, people who shop in local markets in poorer nations experience the greatest exposure to pesticides. It is not surprising, but still shocking, to find that one report concluded that pesticide-caused diseases have overtaken endemic diseases in Third World countries.[52] It becomes increasingly clear that the profit the richer nations of the world receive from the export of pesticides and the accompanying agricultural technology comes at the expense of the health of the citizenry of poorer nations.

The issue of food safety is clearly a serious concern that must be added to the list of negative aspects of pesticide use, along with inadvertent contamination of water supplies and hazards to applicators and field workers. Still, one other hazard of pesticide use—the responses of the pests themselves—has the most direct consequences in terms of farm economies, and is therefore most likely to ultimately lead to a different approach to pest control. This is the subject of the next section.

PESTICIDE RESISTANCE AND OTHER EFFECTS IN AGROECOSYSTEMS

Perhaps the most critical problems associated with agricultural pesticide use are those that affect the fields themselves: the creation of resistant pests, new pests, and intensified pest outbreaks. For though these effects don't directly endanger human lives, they set up agriculture's dangerous addiction to pesticides, creating an ever-increasing perceived need for, and hence increasing use of, these chemicals.

Creating Resistant Pests. In ecological terms, pesticides have created more pest problems than they have solved. Attempts to control 25 major crop pests in California with pesticides gave these results, according to T. H. Coaker's 1977 review paper "Crop Pest Problems Resulting from Chemical Control": 20 of the original pests became resistant to one or more insecticides; in 22 cases, secondary pests became problems after target pests were treated; and nine species, not previously considered pests, reached pest status.

These results can hardly be considered the goal of agricultural pesticide use, yet similar effects have occurred, over and over, throughout the world. These "wonder" chemicals, which some estimate have increased food production by as much as 20 percent since 1940, have also created hundreds of resistant pest species over the same period. By 1986, at least 264 strains of insects and mites, 67 types of plant pathogens, 2 nematodes, and 19 weed species were resistant to pesticides, with numbers still rising.[53] In fact, R. Kowalski notes in "Organic Farming—A Sound Basis for Pest Management?" that crop losses due to pests have changed very little since the introduction of chemical pesticides. It still remains at about 33 percent, with only the source changing: from weeds to insects and diseases. While insecticide use increased tenfold since the 1940s, crop losses to insects doubled.[54] In *Ecology of Pesticides*, Anthony W. A. Brown claimed that no group of pesticides yet discovered is immune from development of resistance, and in fact none is expected to be.

Resistance develops because pesticides never kill 100 percent of a pest population. The survivors tend to be predisposed to a lower susceptibility to the particular chemical used. With every repeated application of that pesticide, these naturally resistant individuals make up a higher percentage of the population, until a highly resistant strain of pest has evolved. This unnatural selection is highly effective and often rapid. Ironically, the more effective a chemical is at killing a pest, the more effective it is likely to be at selecting for resistant pests. A pesticide with a high rate of kill, applied to a pest population that completes several life cycles per season, in a situation where little pest migration from untreated populations occurs, is the perfect recipe for extremely rapid development of pesticide resistance.

Once this experiment in evolution is performed it is essentially irreversible. Scientists do not know how to consciously remove resistance, and nature does not do the job for us. Formerly effective pesticides cannot be successfully reintroduced. Switching to another chemical may help for a short time, but resistance to one chemical often confers resistance to a whole family of related chemicals. Inevitable and irreversible, pesticide resistance is the real enemy in the war on pests.

Creating Superpests. In response to resistance, farmers often use higher and more frequent doses of chemicals, switch to a new chemical, apply mixtures of pesticides, or alternate chemicals at each application. The first two of these tactics speed development of resistance by increasing the kill rate and therefore the selection pressure. The latter two tactics may delay the onset of resistance, but often the price is the development of super-resistant pests—that is, pests resistant to a whole array of chemicals. Such superpests now threaten many crops throughout the world.[55]

In North and Central America, cotton production is seriously threatened by the worst record of superpests of any crop worldwide. The cotton bollworm and budworm are both resistant to chlorinated hydrocarbons, organophosphates, and carbamates, and are fast becoming resistant to pyrethroids. Worldwide, thirty-three species of cotton pests were resistant to pesticides as of 1982, an increase from twenty in 1965.

In Asia, widespread famine would be the likely result if a rice superpest arose. Although none has developed yet, the potential exists in any one of at least fourteen pest species that are already resistant to one or more pesticides. Experts estimate that a lack of pest control could mean a loss of 50 percent of the rice crop in Japan, India, and the Philippines.

In Southeast Asia, brassicaceous crops (cabbage relatives), a dietary mainstay, are threatened by a superpest, the diamond-back moth, which is now resistant to at least eleven insecticides. Resistance in this species extends to all the major groups of insecticides, including the newest introduction, the pyrethroids. In *A Growing Problem: Pesticides and the Third World Poor*, David Bull reports that farmers respond to the threat with applications of multiple chemicals. Up to eleven different insecticides may be used in the course of a season, sprayed as often as three times a week right into the last week before harvest, and all in higher doses than are recommended. These extreme measures highlight both the desperation of the farmers and the hopelessness of the battle.

Creating More Pest Problems. The chemicals used in modern agriculture can have other serious side effects within the agroecosystem in addition to creating resistant pests.

Pest resurgence is one such side effect. Pest resurgence means that nonresistant target pests are killed by pesticide applications, but later the population rebounds to a higher level than occurred before pesticide use. Instead of developing resistance to the chemical, the pests resurge because the pesticide killed the natural enemies and competitors of the pests as well as the pests themselves. When this happens, once the chemical wears off, the environment is actually more favorable to the pest than it was prior to treatment. Without predators, pests can attain larger populations. Many insecticides kill predatory and herbivorous insects alike. In fact, predators are often more sensitive to pesticides than are their prey. This was the case T. H. Coaker described, in "Crop Pest Problems Resulting from Chemical Control," for the predators of the cabbage rootfly: by virtue of its greater mobility, the predator was sensitive to lower doses of pesticide than the fly larvae. Application of aldrin and dieldrin resulted in increased survival of the pest eggs and larvae normally eaten by the predators, and the net effect was intensified crop damage.

Another side effect is the creation of new pests from previously harmless species. This can occur when pesticides destroy the natural enemies or competitors of previously controlled herbivores. Once released from predation or competition, these once-rare species can become abundant enough to damage crops.[56]

Probably the worst case of new pests arising because of chemical treatment of old pests is that involving cotton. Prior to any chemical use, two species were pests, one being the introduced boll weevil. After calcium arsenate came into use (from 1924 to 1945), two different species became major pests and three other pests became problems. In the chlorinated hydrocarbon era, four more species rose to pest status. When organophosphates were added to the pesticide regime, the pest status of one species rose to major status, while two others declined to minor status. The total effect of chemical use was an increase from two major pests to nine.[57]

Sometimes chemicals meant to kill actually stimulate reproduction of pest species. Coaker reported that applications of DDT to beans increased fecundity in one mite species by 70 percent. A follow-up study confirmed increased egg production in this species following exposure to the chemical. A later experiment linked DDT and carbaryl to direct stimulation of reproduction in a mite species when applied at sublethal concentrations.

Some work also indicates that pesticides stress the plants they are meant to protect and make them more vulnerable to herbivory.[58] Herbicides, insecticides, fungicides, and even nitrogenous fertilizer have all been blamed for causing a stress reaction that increases the level of free amino acids and sugars in plant cells. This condition makes plants more appetizing to herbivores and can lead to higher herbivory levels. Studies by Oka and Pimentel summarized in David Pimentel's 1977 paper titled "Ecological Effects of Pesticides on Non-target Species," confirm that herbicide use can stress crop plants and make them more susceptible to pests. They found that corn had greater infestations of aphids and corn borers and less resistance to corn smut and leaf blight when fields were treated with the herbicide 2,4-D.

A further way herbicides have been found to intensify pest problems is simply by working. When an herbicide applied before the crop comes up successfully removes all the weeds, it removes all alternative foods for the resident herbivores. In some cases this heightens the vulnerability of tender, newly emerging crop seedlings.[59]

This problem is of greatest concern when the problem herbivores have generalist feeding habits—in other words, they feed on a variety of plant species.

Creating Addiction. The net result of all these effects of pesticides is an increase in pest problems, an increase in the perceived need for further pesticide use, and a continuous supply of new pesticides. Clearly, agriculture has become addicted to pesticides. While numerous scientists have come to this conclusion, progress toward a different solution is slow (see Chapter 2). What is really needed is a change in perception of the problem. First, it is necessary to recognize that the problems created by killing a target pest population stem from the interconnectedness of the various components of the agroecosystem. Pests do not exist in isolation; they coexist with natural enemies, competitors, and alternate food sources. Second, pests are not a static target, but a dynamic, evolving target capable of creative self-defense. Attempts to deal with this complexity by simply ignoring it, or worse, eliminating the complexity by simplifying the agroecosystem, continually meet with failure. Until modern agriculture learns to work within the complexity nature provides, it will be difficult to move away from chemical addiction toward a more sustainable solution to pest problems.

2

Roots of the Crises

THE TWENTIETH CENTURY has seen revolutionary changes in the way agriculture is practiced. Farming has moved from a cultural art, handed down from generation to generation, to an industry, taught as a profession by expert specialists who may never have farmed a day in their lives. Along with the rise in production that accompanied the change to industry have come the environmental problems of industry, a drastic reduction in the number of farmers, and economic instability.

As the word itself implies, agriculture is a part of culture, and the roots of the current problems in agriculture can be traced directly to basic cultural values. Both scientific and economic choices reflect these values, though they are rarely acknowledged outright. This chapter traces the roots of the problems described in Chapter 1 to the different phases of agricultural industrialization. It explores the scientific setting of agriculture for the philosophical and practical reasons behind the choice of specific research problems and the types of solutions obtained. Next comes an examination of cultural values that have shaped the modern perspective on agriculture and produced the current unsustainable system. Finally, the discussion moves to the concept of sustainability, what this means in agricultural systems, and what changes in cultural values and in scientific and economic choices are necessary to accommodate a goal of sustainable agriculture.

ROOTS OF THE CRISES IN AGRICULTURAL INDUSTRIALIZATION

The serious ecological problems caused by modern agriculture are the often overlooked aspect of agricultural industrialization in the twentieth century. They are a product of the revolutionary changes toward industrialization of agricultural practice in this century. The economic instability of farms is rooted in the process of industrialization also. The key components of this industrialization process—mechanization, chemicalization, and new crop-breeding priorities—have each solved certain problems that previously limited agricultural production, but, at the same time, each has created new ecological and economic problems (see Table 2.1).

TABLE 2.1

Modern Agricultural Practices That Have Contributed to the Current Ecological and Economic Crises

Practice	Problems addressed	Problems created
Mechanization	Labor inefficiency	Erosion, energy dependency, capital expenses, interest payments, larger farms, fewer farmers
Inorganic nitrogenous fertilizer	Crop yield	Groundwater contamination, farm specialization, pests, erosion, energy dependency, high input expenses, less economic resilience
Pesticides	Crop loss to pests (success doubtful)	New pests, resistant pests, water pollution, human poisoning, energy dependency, high input expenses
Hybrids and genetically narrow varieties	Crop yield and nonuniform traits	Aggravated pest problems, loss of local adaptations, chemical dependency, high input expenses

MECHANIZATION

As with the Industrial Revolution in general, the agricultural revolution is most obviously characterized by the process of mechanization. This took place in two phases: the first phase from about 1820 to 1870, when animal power replaced human power, and the second phase between about 1920 and 1950, when the gasoline engine replaced draft animals. With mechanization, a farm family's labor was no longer the limiting factor for how large a farm could be. Farms began to expand because of the extra time available to farm more land. This process set in motion the ongoing drastic reduction in the number of farms and the simultaneous rise in average farm size. Today, all agricultural activities, from plowing to threshing and beyond, are mechanized on the typical modern farm. Fossil fuels and electricity are substituted for labor in every imaginable way. In the process, mechanization has contributed to many of the problems described in Chapter 1.

Without a doubt, mechanization has been highly efficient in increasing the labor efficiency of U.S. agriculture. Today, the United States produces more food per unit of labor than anyone else in the world. Yet mechanization has had many effects that run counter to the sustainability of agriculture. On the negative side, mechanization has reduced the use of conservation measures and led to more soil erosion, more fossil fuel consumption, and greater debts and slimmer profit margins for farmers.

Mechanization has contributed to soil erosion by several routes. One route is through changes in field size and layout. Large equipment is easier to operate on large fields, following the field's geometry rather than the land's contours. But larger fields have fewer windbreaks and therefore greater vulnerability to wind erosion; and geometric, rather than contour, plowing exacerbates water erosion.

A second route involves physical changes in the soil caused by machinery. Heavy machinery compacts the soil, which reduces the soil's moisture-holding capacity and the rooting depth of crops. Both factors can accelerate erosion. In addition, high-powered, rubber-tired tractors plow at higher speeds, which causes more tillage erosion.[1]

A third route is through the changes in cultural practices brought about by mechanization. Machine traction contributed to the removal of animals from most cash-grain farms and also meant the removal of draft animals from farms. Before mechanization, animal manure had been an important source of organic matter returned to the soil to help reduce erosion. Furthermore, expensive, specialized equipment requires continual production of the crop it was designed for and so contributes to the demise of crop rotations, which can reduce erosion in some cases. Finally, and more subtly, machinery has worked to disguise erosion so that it can be ignored. More powerful tractors can wipe out telltale physical signs, such as rills and beginnings of gullies, and can handle heavier soils as topsoil depth diminishes.[2] They provide the power to till poorer, heavier, more vulnerable soils, soils that should not be plowed in the first place.

Mechanization has also contributed to agriculture's energy dependency. As of 1973, the manufacture, operation, and maintenance of farm machinery accounted for the majority (51 percent) of worldwide agricultural energy use. In North America, mechanization used more energy, close to 60 percent.[3] These figures include manufacture and maintenance energy use, in addition to fuel use. Machinery is especially problematic as we approach the limits of fossil fuel supplies, because it depends on liquid fuels and is not easily converted to renewable sources, such as solar or wind power. It is a key aspect of vulnerability due to energy dependency.

Some of agriculture's economic woes began with the process of mechanization. The costs of farming, the debt load, and hence the interest load rose as expensive machinery became a necessity for competitive farming. Loans for the purchase of machinery are a major debt burden on many farms. A four-wheel-drive tractor with air-conditioned cab carries a price tag four to five times that of a family car. Prior to the reforms of 1986, tax laws permitted fast depreciation and favored replacement of equipment before its useful life was up. As bigger, more powerful equipment became available, farmers found they could farm more land in less time, so possession of the machines became the motivation to increase farm size (and thereby increase debts on land). This set in motion the expansion of farms that has emptied the rural countryside and is disrupting rural economies. Economic vulnerability due to specialization on a single crop is partly a result of specialized, expensive machinery that must be used frequently to justify its costs.

Finally, giving farmers the means to farm more land in less time sets in motion what Willard W. Cochrane in his classic 1958 book *Farm Prices: Myths and Reality* called the "agricultural treadmill." Since the prices farmers receive for their crops are dictated by the total amount of a crop produced by all farmers, each farmer has no control over the prices he or she receives. When a few farmers adopt a new technology that increases their output, the prices received by farmers are lowered for everyone. Pressure builds for all farms to become bigger or increase efficiency. In evolutionary biology, a similar phenomenon is called the Red Queen hypothesis (referring to what the Red Queen said to Alice in *Through the Looking Glass*): you have to keep running faster and faster just to stay in the same place, since all species are continually becoming better adapted to their environment. As long as farmers' rewards are determined by their production, this hypothesis holds true. If farmers could be rewarded in some way for their stewardship of the land, or for the quality of their products, or for the thriftiness of their operation in terms of charges against nature and future generations, they could step off the agricultural treadmill and progress toward a more sustainable system.

CHEMICALIZATION

After World War II came another revolution in agricultural practices, one that could be called chemicalization. It consisted of the substitution of chemicals for organic fertilizer, mechanical tillage, and cultural methods of insect and pathogen control. These changes simplified and standardized farming so that cultural wisdom no longer appeared to be essential. Formulas and instructions from chemical companies could be read like a recipe book, with the farmer's intellect and experience becoming less valued.

Inorganic nitrogenous fertilizer became available at cheap prices following World War II, when the industries that had been producing explosives found themselves without a market. Timing was ripe, as soil erosion had been taking its toll. Today, nitrogenous fertilizer use continues to increase, causing groundwater pollution. It also contributes to soil degradation in several (some subtle) ways, so that it sets up the addictive cycle—the more used, the more needed.

As described in Chapter 1, nitrogenous fertilizers disguise soil erosion at the same time that they exacerbate it. Adding nitrogenous fertilizer reduces the organic matter in soil by stimulating decomposition, and this in turn changes the physical properties of the soil that influence moisture-holding capacity, compaction, and vulnerability to erosion. Ironically, the inorganic fertilizer added to replace lost soil fertility ends up promoting processes that further reduce the soil's fertility and demand further fertilizer additions. Yield losses due to soil loss have so often been disguised by yield increases due in part to added fertilizers that soil erosion has been overlooked. This has created a mind-set that minimizes the importance of conserving our soil.

Chemical fertilizers also contribute to the energy dependency of agriculture, as their production is highly energy-intensive. In the United States, fertilizer accounts for some 30 percent of farm energy use.[4]

In terms of economics, the availability of cheap fertilizer facilitated the trends set in motion by mechanization and thus has contributed to the loss of economic resiliency, the exit of animals from farms, and the abandoning of crop rotations with legumes. One chronic source of debt is production loans to pay for purchased inputs: fuel, seed, fertilizer, and pesticides. In recent years, on average, farmers have had to purchase 80 percent of their production items and have produced only 20 percent on their farms. Usually the purchases have required short-term operating loans. Those purchases, excluding interest on loans, ate up some 52 percent of farmers' cash income from farming in 1982, up from 38 percent in 1950.[5] Prices of fertilizer, fuel, and pesticides all fluctuate with oil prices, so there is always uncertainty in the equation for farmers, except that oil prices will inevitably rise with scarcity. If farmers buy inputs on credit, they are at the mercy of interest rate changes each year. Thus, reliance on manufactured chemical control of fertility and crop protection contributes to financial vulnerability.

New chemical pesticides were developed after World War II as an offshoot of chemical warfare. As already noted, the chemical war on pests has in many ways proven a miserable failure, creating worse pest problems than existed before the introduction of chemical pesticides. This quick technological fix has proven unequal to the myriad dynamics of nature. And, although pesticides have provided many short-term rescues from pest problems, in the long run they have not dramatically reduced overall pest damage.

Like machinery and fertilizer, pesticides add to the energy and input costs of farming, thus increasing energy dependency and financial vulnerability. The trend toward specialized farms and widespread use of genetically narrow varieties, and away from crop rotations on farms, all have made crops much more vulnerable to pest problems. At the same time, the economic consequence of these trends—reduced economic resiliency—has reinforced dependency on pesticides' false promise of sure results.

CROP-BREEDING PRIORITIES

The third key component in the industrialization of modern agriculture involves changes in crop-breeding strategies. Modern breeding techniques have focused on two goals: higher yields and increased crop uniformity. These goals parallel the industrial goal of standardization to improve production and efficiency. Techniques that greatly restrict the genetic diversity of crops have helped achieve these goals and have also contributed to a loss of breeding material, as noted in Chapter 1. Hybrids formed from crossing two inbred (genetically uniform) lines have been the centerpiece of crop breeding in cross-pollinated crops. Corn yields have indeed risen fivefold since hybrid corn was introduced in the 1930s,[6] but not all the increase was due to hybridization. Some was due to fertilizer use and some to selection within the inbred parent lines. The hybridization technique has several consequences than can be seen as either negative or positive, depending on one's viewpoint.

One consequence is uniformity. A field of hybrid corn has very little variation among individual plants. Uniformity of expression of such agronomic traits as height and ripening date aids farm operations and improves the harvest, but it also allows crop pests to adapt quickly to exploit the crop. This danger has been acknowledged by corn breeders in recent years, and techniques that increase genetic variation have been introduced into breeding programs.[7]

Another consequence of the hybridization technique is that hybrid plants do not produce seeds similar to the seeds from which they grow. This means that farmers cannot save a portion of the crop for seeds to plant out the next year, but must purchase new seed each year. The benefit of this consequence goes to the seed companies. It accounts for the great interest of seed companies in crops that can be hybridized. It has no doubt contributed to the speed with which improved varieties have been produced, because of the clear potential for financial gain. The disadvantage of this aspect of hybridization goes to farmers, who must add the price of seed to their annual production costs.

The need for hybridization to achieve higher corn yields has been questioned.[8] Some argue that corn genetics indicate that even greater gains could be achieved by mass selection. This technique would not reduce genetic diversity to the degree that hybridization has and would produce open-pollinated plants with savable seed. In other words, farmers would not have to purchase seed each year. In addition to the monetary savings it offers, saving seed can be a benefit in that it is a form of selection itself and can work to adapt a variety further to the local conditions where it is grown.

While the genetic basis of the argument against hybrid corn is controversial, an important concept is involved. The suggestion that hybridization is not necessary raises the issue of whether agricultural research has taken modern agriculture where society wants it to go. Was developing hybrid corn the best path to take, or would it have been better to work toward open-pollinated varieties, even if it had taken longer to achieve equivalent yields? In their chapter in *Agroecology*, Richard C. Lewontin and J. Barlan posed the question: "What if agricultural research had devoted *equal time and energy* to a different goal? Where would we be now?" They ask citizens to question profit-motivated research and ask if it has truly resulted in the best agriculture possible.

Another controversial aspect of modern crop breeding is the extent to which it has been done in a chemically controlled environment. How much has the use of inorganic fertilizers and chemical pesticides influenced the apparent need for chemical inputs in the field? If the crops had been bred in a low-input environment, would there now be crops that thrived with fewer inputs? These questions are valid and need to be given serious consideration.

In sum, because an industrial model has been applied to a biological system, myriad biological problems have developed in modern agriculture. As solutions have been sought within that model, ecological and economic problems have become intertwined. To get past the current crisis juncture, today's agriculture needs revamping, both ecologically and economically, from the soil up.

ROOTS OF THE CRISES IN AGRICULTURAL SCIENCE

The industrialization of agriculture came about through a conscious process of applying mechanical ingenuity and scientific methods to agriculture. It took place through carefully designed experiments in which many factors were controlled and a very few responses were observed. This was a new approach. Prior to the late 1800s, agriculture had been a cultural art, improved through direct experience and trial and error. With the birth of agricultural science, it was to become a profession. The scientific approach provided solutions to many technical problems and, as noted earlier, permitted many other problems to be overlooked. An examination of the institutions that have produced this scientific transformation helps clarify the roots of today's dilemma.

ROOTS IN THE INDUSTRIAL REVOLUTION

Our agricultural science institutions are now and have always been closely aligned with industry. From their inception, the public agricultural research institutions in the United States have approached their work from the vantage point of the industrialist entrepreneur. This view dictates that solutions be compatible with the goal of maximizing profitability via maximum production and minimum labor. It follows from the premise that the country has plentiful resources and a shortage of labor.[9] In 1862, when the foundations of these institutions were laid, farming was in the process of transformation from subsistence-type frontier farming to market production. The advent of transcontinental railroads facilitated this change, and growing urban populations and surplus agricultural production made change a necessity. By 1870 farmers were beginning to see farming as a profession instead of as a way of life and were actively seeking help in improving their profits.

Establishing the Institutions. The year 1862 was a landmark year for American agriculture, for within a two-month period President Lincoln signed four separate acts that were to have profound effects on agriculture: the Act to Establish a Department of Agriculture, the Morril Land Grant College Act, the Homestead Act, and the Transcontinental Railroad Act.

The initial purpose of the United States Department of Agriculture (USDA), according to the establishing act, was to "to acquire and to diffuse among the people of the United States useful information on subjects connected with agriculture in the most general and comprehensive sense of that word, and to procure, propagate, and distribute among the people new and valuable seeds and plants." The first commissioner, Isaac Newton, set the direction of the department toward scientific research and education. In the 1880s, the USDA also began to acquire regulatory responsibilities, starting with testing butter for adulteration.

The land grant colleges of agriculture were not formally connected to the USDA, but since most USDA employees were soon graduates of these colleges, the two systems were informally connected. The initial purpose of the land grant colleges was to provide a "liberal and practical education of the industrial classes." They were to serve those who could not qualify for or afford private colleges, and the act was partly an attempt at Jeffersonian-style democratization of education. But quickly the practical part of this charge came to the forefront, and many saw the primary purpose of these colleges as providing education in improved methods of agriculture. This created the dilemma of a shortage of information to teach. Farmers were losing patience with the colleges; the colleges needed research results.

The upshot was the establishment of the State Agricultural Experiment Stations (SAES) in 1887 with the federal appropriations authorized under the Hatch Act. The purpose of these stations was to conduct agricultural research appropriate for each state. They were mostly located at and associated with the land grant colleges and were largely independent of the USDA. The act stated clearly that the purpose of the experiment stations was to conduct "researches, investigations, and experiments bearing directly on and contributing to the establishment and maintenance of a permanent and effective agricultural industry." The goal of the research was to obtain "useful and practical information on subjects connected with agriculture, and to promote scientific investigation and experiment respecting the principles and applications of agricultural science." In 1914 the final piece of the puzzle, the Cooperative State Extension Service, was established with the Smith-Lever Act. It was to be a joint effort between federal (USDA) and state (SAES) research institutions to take research results to the farmers and to bring farmers' needs to the attention of researchers.

Whereas the initial wording of the various acts establishing our institutions of scientific agriculture did not preclude—in fact, even demanded—the establishment of a "permanent" agriculture, this did not mean quite the same thing as we do when we advocate a "sustainable" agriculture. The scientific agricultural institutions were born in rhetoric promoting the good of agriculture, with social welfare the aim, yet they were also born in an atmosphere in which the

nation romanticized industry. In this context, "advances" that bene-
fited some farmers at the expense of others, who as a result were
driven out of business, were all for "the good of society." The ex-
farmers were then "free" to take more productive, higher-paying jobs
and to pursue "the good life" in the cities. Whereas the initial charge
to the experiment stations talked about a permanent agricultural industry, the emphasis was on industry rather than farming. At that
time, those words meant something closer to "figure out how the
most people (not just farmers) can make the most money in the process of providing the urban public with food."

Roots in the Patent Office. A look at the historic impetus for the USDA
shows more clearly the roots of agricultural industrialization. The
major impetus for a federal Department of Agriculture came from a
coalition of agricultural societies composed of wealthy intellectuals
who wanted the United States to keep up with advances in European
agriculture. Doctors, professors, politicians, businessmen, and some
gentleman farmers formed their ranks. These societies tended to be
centered in cities, and, in fact, a requirement for full membership in
the prestigious Philadelphia Society was to live with a ten-mile radius
of the city. This particular group's constitution stated that its purpose
was to promote "greater increase of the products of the land"—in
other words, higher yields.

Prior to these groups' lobbying efforts, one of the most vociferous
advocates for the concept of a USDA was Henry Leavitt Ellsworth,
the commissioner of patents. He was faced with such a deluge of new
designs for agricultural machinery and such an influx of improved
seed varieties, that he believed a government department was
needed to test inventions (to protect farmers from fraud) and to develop and distribute improved seed varieties. Thus, the beginnings
of the science of crop breeding and the first phase of agricultural
mechanization from 1840 to 1900 were entirely the product of entre-
preneurial capitalism. Agribusiness preceded agricultural sciences in
this country.

Ellsworth saw clear potential for enhancing the country's economic
growth by increasing agricultural production and promoting the
growth of the farm machinery industry. Writing in the 1830s, he rea-
soned that if the country then consumed $120 million in wheat prod-
ucts, a 10 percent yield increase due to superior varieties would
translate to an additional income generation of $15 million to $20 mil-

lion for the country. He had complete faith in the expansion of market demand to meet the increased supply. This principle guided the country into surplus grain production by the early 1900s. Despite an ongoing surplus and depressed prices ever since (except during wartime), the goal of increased production never slipped from the top slot of U.S. agricultural research, and Ellsworth's faith remains alive today. It is clear in this 1985 statement by a USDA assistant secretary: "Thus, surplus is basically a marketing problem which research attacks by improving product quality and reducing product cost . . ."[10]

CONFLICTS BETWEEN SCIENCE FOR FARMING AND SCIENCE FOR AGRIBUSINESS

Using science to improve agriculture produced dramatic results. Some of the early accomplishments of the USDA included introduction of the Australian vedalia beetle (ladybird) to control cottony cushion scale (also Australian) on California citrus trees (originally from Asia), breeding disease-resistant strains of cotton to combat wilt disease, and development of a butterfat test for milk products for quality control. While these accomplishments were fairly egalitarian in effect, other changes were much less so. Growth in productivity in the agricultural sector was much more dramatic in agribusinesses associated with farm inputs and the processing of raw farm products than in income generated by farmers.

What was neatly overlooked by many was that what was good for agribusiness was not necessarily good for farming. Farm-input manufacturers' profits have often come at the expense of farmers' profits, or at the expense of farmers' options to choose ecologically sound methods of farming. Only 10 percent of the value added in agriculture is added on the farm. In other words, about 10 percent of the price paid for food reflects the work of the farmer. Forty percent is added in creating inputs that the farmer must purchase, and 50 percent is added between the farm gate and the table.[11]

What is good for farm-input manufacturers is farmers who continue to need their products. It is good business for pesticides to be addictive, but it is not good farming practice. It does not help the farmer make the budget balance, and it does not help keep the farm and surrounding neighborhood a healthy place to live.

It is good business for a seed company to produce seeds that do not breed true, so that they must be purchased each year. It does not help farmers to have to add seeds to their list of purchases for the year.

It is good business, when chemical companies own the seed companies, to produce varieties that thrive with heavy fertilizer applications and pesticides. But farmlands and farmers would be better off with varieties that thrive in soils rich in organic matter, are competitive against weeds, and are mixtures of lines with a variety of resistances, so that pesticide use could be minimized.

It is good business to keep inventing more expensive and more sophisticated machinery so that farmers have to continue upgrading to keep up with the competition. But farmers who can operate thriftily by repairing and altering old equipment as needed have one less source of major debt.

It is potentially very profitable for a company to develop a crop that resists herbicides so that it can be grown using herbicides and marketed as a seed-plus-herbicide package, as some biotechnology companies are now pursuing. Though this may be useful for soil-saving reduced tillage, over the long run it will not help farmers or the general public for herbicide exposure to rise, nor does it help for seed prices to rise.

Clearly, the research interests of farmers, the public, and agriculture-related businesses do not always coincide. But it is difficult for agricultural scientists to hear the protests of farmers and the public above the rabble of enthusiasm coming from agribusiness. It is hard to argue with success, and the dramatic increases in agricultural productivity credited to agricultural science appear to be proof of the virtue of the system. The problem here is that only the volume of change is being evaluated, not the direction of change. The question must now be asked: Has agricultural science created the sort of agriculture desired, considering factors other than simple productivity? Society might wish for a more permanent, ecologically sustainable agriculture, or one that is more congenial to young farmers, or one that supports more viable rural communities, for example.

Even more critical, it is necessary to ask whether one of these different kinds of agricultural systems could perhaps achieve productivity equal to current productivity, given equal time, energy, and money devoted to research and development. The controversy over hybrid corn described above centers on this question. Researchers must be cautious about successes and not automatically assume that success in one arena means the best overall solution has been found, or even that the right questions have been asked. In the case of corn, it is possible that farmers could have varieties more compatible with the goal of sustainability if researchers had followed a different path and aimed at developing high-yielding varieties with savable seeds. What sort of productivity might exist in the Corn Belt today if researchers had spent the last fifty years concentrating on developing varieties that would grow best with a groundcover legume that would both hold the soil in place and provide nitrogen? In evaluating the improvements in agriculture, agriculture's current state should not be judged against its beginnings so much as against alternative end points, given a different research focus for agricultural "improvement."

AGRIBUSINESS AND AGRICULTURAL RESEARCH

It is not so much a matter of agricultural research institutions betraying the public trust by aiding in the tremendous growth of the entire agricultural sector at the expense of many farmers. Rather, these institutions have done a fine job of fulfilling a shortsighted, misguided, or at least outmoded trust. After all, in a backward way, it is easy to argue that science has fulfilled farmers' needs. When farmers must sell products to large capital-intensive firms, the firms dictate what the farmer needs. If the buyer wants hard tomatoes, the farmer *needs* hard tomatoes, and the research institutions provide them with just that. For crop varieties, institutions often do the long-term work of developing desirable lines, and the private companies do the final development into a product, receiving the patent rights and profits.[12] As one commodity group leader said, "Let's don't kid ourselves. Our industry is a shining example of the success of agricultural research."[13] How has private industry managed to get so much out of our public institutions?

One route is by private companies providing funding for public research for the initial steps in plant breeding, for pesticide testing, and for machinery engineering and then receiving, in turn, results that can be developed into profitable products. From the mid-1950s to the mid-1980s the portion of funding for all agricultural research, both public and private, coming from private industry rose from about one-third of the budget to half to two-thirds of all funds. An industry-supported survey described by Edwin Crosby found that the industry-supported research budget was $1.7 billion to $2.6 billion in 1985, compared to $1.7 billion in public funding.[14] This shift toward private domination of agricultural research funding occurred as federal allocations dropped or did not keep pace with inflation. This support is not just charity. Companies investing in agricultural research see great potential for profit and expect to reap the benefits. When private funds are used, it is easy for scientists to come to see private enterprise's goals as their own goals.[15] Thus, herbicide-resistant crops (to facilitate more herbicide usage) receive higher priority than competitive crops (to better tolerate weeds and reduce herbicide dependence) or cropping systems that minimize weeds via groundcovers and rotations.

Although total private spending on agricultural research is very large, the amount provided to public institutions represents a small percentage of the total dollars spent by companies on agricultural research, generally about 5 percent. Most privately funded research is conducted in house within private industries and is devoted to product and market development. Yet the small amount of money going directly to public institutions has a greatly disproportionate influence on scientists and research agendas. A study of the Department of Animal Sciences at the University of Nebraska by the Center for Rural Affairs[16] found that private funds often pay graduate student stipends and provide funds with few strings of accountability attached for research support. This helps scientists pay their own way and increases their clout and access to departmental discretionary funds. "In short, the private grants leverage public funds."[17] Most of the overhead costs of the research for which private funds were granted are paid out of public funds—facilities, equipment use, and clerical support services all come from formula funds allocated by federal or state governments. The private funds pay for the extras: computer time, graduate student labor, etc. In the end, Marty Strange says in *Family Farming*, scientific inquiry becomes "the handmaiden of private interests. It reduces scientists to grantsmanship opportunists. Most important, it thwarts complete discussion of the public interest in choosing among alternative paths to technical development."[18]

Given the initial roots and goals of our agricultural research institutions, the partnership between industry and public research is entirely natural. Whether or not this partnership is just or proper, and whether it has produced the type of agriculture needed for the future, is the subject of the current debate. The authors' stand here is clearly that it is neither just nor adequate given the current state of affairs. It was part of the original premises that agricultural research would increase the profits to be made in the entire agricultural sector, not just in farming. However, it must be pointed out that now, given the knowledge available about the ecological and economic consequences of modern agriculture, the original premises of agricultural research no longer make sense. Expanding the profits of the agricultural-input industries is no longer an appropriate focus. What is needed is to move beyond the search for *more* and begin a serious search for *permanence*, for sustainable production in a world of diminishing fossil fuels and increasing ecological stress. And it is quite likely that in order to achieve this goal the current public/private partnership in research will no longer work.

BARRIERS TO CHANGE IN AGRICULTURAL RESEARCH

Agricultural research institutions' preoccupation with expanding economic yields was a product of a pioneering society. Affluence, abundant resources, new land for expansion, exploitation, and mining characterized the period in which these institutions developed. This was the colonizing phase of our history. Now, as resources— including soil, energy, and clean water—approach critical lows, and unemployment and poverty become greater social problems than scarce labor and food shortages, it is becoming clear that we must seek an equilibrium state, a steady state with our environment. As the "public good" is redefined, the research community must also redefine its goals to meet the "public good." But there is a great deal of inertia in the system that locks programs and viewpoints and ways of solving problems in place.[19] The institutions that developed during the colonizing, industrializing period of our history may have little to offer the next phase of agriculture unless a major reshuffling can be achieved. Following is a brief review of some of these barriers to a change of focus toward sustainable agriculture within the research institutions.

Careerism and Institutional Inertia. One source of inertia is the professional reward system for scientists in the universities. Agricultural research was originally mission-oriented, with research topics dictated by administrative priorities and publication of results directed toward farmers and other users through experiment station bulletins. Recently, agricultural disciplines have become much more like the traditional scientific disciplines, and the traditional ingredients of professionalism—publication in disciplinary journals and the acquisition of grant money—have come to have great importance. Competitive grants are a recent introduction to agricultural research. Traditionally, funding came primarily from formula funds allocated by governments on the basis of state rural population. In the 1977 Farm Bill, a small competitive grants program that accepted proposals from both land-grant and non-land-grant researchers was begun. Formula funds decreased in the 1980s and competitive grants increased, though they are still a small source. Competitive grants supplied only $16.5 million to agricultural research in 1982, compared to $140 million in formula funds dispensed by the USDA to the SAES and an additional four times that amount from state formula funds.

After extensive surveys of agricultural scientists in the United States, Lawrence Busch and William Lacy concluded in *Science, Agriculture, and the Politics of Research* that research is now primarily dictated by the scientists' own interests and perceptions of the needs of agriculture. Scientists rarely consult with farmers concerning research needs. Competitive grants from federal agencies tend to lock a researcher into past channels. It is very difficult to get funded for a pilot study for a new project that differs much from one's previous research. Therefore a researcher gets locked into a narrow range of research choices that all relate to one another and draw on previous experience. These projects are less risky for funding institutions, which consider them more likely to produce clear results because the bugs of techniques have already been worked out. Prestigious positions on grant review boards tend to go to older, established scientists, who are not likely to fund "interesting" innovative research that may fly in the face of their own life's work. But the process severely limits innovation in addressing new problem areas or trying new approaches.

Formula funds received from state and federal appropriations also tend to be distributed to existing research programs. Administrators tend to go for program maintenance, to hang on to the tried and true.[20] Thus, the different land grant universities come to be associated with particular research and to specialize in agricultural products especially important in their region. Kansas State University is the world center for research in wheat breeding, milling, and baking, while the University of Nebraska Animal Science Department is a leader in animal nutrition. While regional specialization is appropriate for state experiment stations, institutionwide specialization means that new concepts or innovations must compete with a huge investment in the status quo. Consequently, some constituencies have very little chance of being heard.

Peer-reviewed disciplinary journals also tend to be a conservative force in agricultural research. Journal editors act similarly to grant reviewers by weeding out new approaches. Evidence of this narrowness is found in the tendency for new journals to appear when new topics emerge. This is often the only way for innovators to get published. Following the energy crisis of the 1970s a number of new journals appeared, including *Applied Energy, Biomass,* and *Energy in Agriculture.* The latter published only six volumes before it folded in 1988 as interest in energy issues waned. Most recently, *American Journal of Alternative Agriculture, Biological Agriculture and Horticulture, Agriculture, Ecosystems and Environment,* and *The Journal of Sustainable Agriculture* have appeared so that research on sustainable agriculture has places to be published.

Because tenure and promotions are largely based on ability to attract grant money and on publication records, it is natural for researchers to choose projects on the basis of where and how fast results can be published. Funding periods for grants are typically short also, on the order of two or three years between renewal attempts, and even formula funds are subject to frequent appropriations battles in state legislatures. All of this adds to the pressure to publish and biases research toward short-term gains rather than long-term solutions. It is little wonder that long-term consequences of many agricultural practices were overlooked for so long.

The combination of pressures to conform from granting agencies, journals, and administrations in order to survive and advance in public research institutions not only limits the problems addressed, but limits the solutions sought as well. The response of the agricultural research community to the need for nitrogen fixation research to reduce reliance on inorganic nitrogenous fertilizers is a good example.[21] A 1977 Office of Technology Assessment report summarized the possible routes for such research. One involved improvement of the traditional legumes-in-rotation scheme by the selection of nonconventional native species with nitrogen-fixing ability. The other path involved biotechnology, either to transfer rhizobial nitrogen fixation from legumes to other crops, or to transfer the microbial process of fixation directly to higher plants. Although legume rotations were tried and true (though not much researched and improved), and the biotechnology solution completely untested, the latter better fit the current paradigm and received the most attention in the report, as well as by far the most subsequent research attention. Most biotechnology during the 1980s was funded privately. The USDA had funded neither type of research as of 1983, but it has since begun to get involved in biotechnology research. In the 1988 list of experiment-station research priorities, biotechnology was ranked second in priority and far and away first in recommended level of funding.

It is not surprising that biotechnology won out, even though it was an innovative and untested approach. It is more formula-oriented, promotes uniformity in agriculture, requires sophisticated science rather than farmers' common sense and field trials, and most important, has the potential of producing patentable, profitable products. It fits the paradigm of humans controlling, altering, and improving on nature. In contrast, legumes in rotation with other crops fits a paradigm of humans working with nature, and seeking ways to make better use of natural elegance, rather than trying to improve upon it.

Communication Barriers. Another barrier to the development of sustainable agriculture within our public research institutions is the difficulty of interdisciplinary communication. Soil scientists, plant pathologists, entomologists, economists, and rural sociologists, like all other scientists, tend to develop their own terminology. Communicating across disciplines becomes akin to communicating in foreign languages. The result is that research coming out of one discipline may be counterproductive to the work of another. For example, while crop breeders concentrate on maximum yield, they may inadvertently create new insect and pathogen problems that the entomologists and plant pathologists must tackle. What is needed are cooperative efforts to achieve simultaneous optimization of as many important factors as possible. If crop breeders, plant pathologists, and entomologists worked together, or at least worked out common goals, such conflicts could be avoided and some progress might be made toward sustainability.

One of the most curious examples of disciplinary isolation in our universities is the separation of ecology from agriculture. Ecology is the study of interactions among organisms in ecosystems and of ecosystem processes. Farm fields are certainly ecosystems, but agricultural research has often viewed them more as factories and has largely been directed toward making them function as such. Until recently, there has been little blending of ecology and agriculture, but with the advent of the USDA competitive grants program in 1977, just when the field of ecology was filled to overflowing, ecologists began to see agriculturally altered ecosystems as valid research sites. Although this is a backward way to infuse an ecological viewpoint into agriculture, it is nevertheless bringing together some ecologists and agriculturalists on interdisciplinary teams. In the 1960s and 1970s, British ecologists led by John L. Harper reached into the agriculture literature and found studies that led to an explosion of population-based plant ecology studies and theory on competition and other interactions among plants. This is part of the reason for today's interest in agriculture by ecologists, who are motivated by the elegance of a simplified ecosystem for working out basic ecological relationships. Now may be the time for a return flow—with ecological research providing insights into how to take advantage of natural processes and relationships in agroecosystems to reduce reliance on fossil fuels and chemicals and reduce negative environmental impacts.

Finally, the isolation of the disciplines limits the range of solutions to problems. We can see this in the process of translating a farmer's question into a research project. A question such as "Why has this corn died?" has many levels of answers. It could be answered by describing the disease and noting the mechanism of cell destruction (a plant pathologist's viewpoint). It could be answered by noting that the variety of corn planted is susceptible to the disease contracted and that the weather was just right for the disease this year (a microbial ecologist's or breeder's viewpoint). It could be answered by pointing out that the field has been in continuous corn production for the preceding ten years, the soil was poor in humus, and the disease has probably been building for that whole period, just ripe for an epidemic as soon as the weather was right (an ecologist's viewpoint). The solutions could range from a prescription for a fungicide application, to a new resistant variety of corn, to a crop rotation scheme that included multilines of corn with a variety of disease resistance, a plan to build soil humus, and perhaps a windbreak to prevent spread of disease from a neighbor's continuous-corn field. The array of responses would depend on what types of scientists were asked the question and what their training had been. The implications of the different solutions are profound indeed. As long as the goals aren't clear,—is the cheapest solution desirable? the quickest? the most long-lasting? the one with the least environmental impact?—we may miss the most appropriate solution. The difficulty of interdisciplinary cooperation in searching for solutions nearly ensures that we will miss it.

With the overriding mission of agricultural research defined as efficient, maximum production, it is difficult to achieve progress toward sustainability. Efforts are only piecemeal and may often contradict one another. For example, no-till cultivation prevents soil erosion but endangers groundwater quality. Given the initial assumption that agriculture was an industry open to improvement by mechanization and application of science to increase yields, it is not surprising that researchers have rarely addressed wider ecological problems until crises have occurred. Currently, coordination among departments and subdisciplines is poor, and common goals are vague. The research goal still most widely shared is increased production, not sustainable production. The good news is that with so little time, energy, and research spent on sustainable practices to date, the room for improvement, given a little attention, is surely vast.

ROOTS OF THE CRISES IN UNDERLYING VALUES

So far this chapter has explored how the application of industrial techniques and economic goals to agriculture has produced the modern agricultural system—highly productive, but fraught with serious problems that threaten its sustainability. It has been explained that today's agricultural research system is unlikely to take the lead in creating a sustainable agriculture because of built-in barriers to innovation and its intimate connections to business interests that run counter to sustainable practices. But the values and assumptions that underlie the research system have not yet been examined, and these are fundamental to understanding why it has been so easy to ignore the impermanence of many solutions to problems and the unsustainable nature of modern practices. This section takes up that exploration.

SCIENTIFIC VALUES AND ASSUMPTIONS

At the foundation of modern science is the concept of reductionism. The tendency to find a minimum set of factors (preferably one) that appear to have controlling influence on a particular outcome characterizes the reductionist approach to science. The desired outcome is usually reduced to a single, easily measured factor. Underlying reductionism is the belief that the whole can be understood as the sum of its parts. In other words, the behavior of each component independently will predict the behavior of the whole. The weakness of this approach is that interactions among components, side effects, and long-term consequences are often overlooked. For example, a study of the effects of nitrogenous fertilizer on a crop might concentrate on yield effects and miss such indirect effects as changes in rates of organic matter decomposition or changes in the crop's palatability to herbivores. These subtle effects could have long-term consequences.

Simplification. The reductionist approach is the legacy of seventeenth-century philosophers, especially the mathematician René Descartes, who gave us the Cartesian coordinate system (the two-factor graph). This approach has its benefits. It allows us to discover critical factors—those factors that have a strong influence on a system. It helps us to pick out patterns within great complexity. But to the extent that it narrows the set of causal factors considered, and, even more, the set of effects observed, it opens us to the danger of missing a great deal of the functioning of complex systems, such as agroecosystems.

Especially vulnerable to neglect are the small but cumulative effects that may take many years to be noticed if not actively sought out. Nitrogen contamination of groundwater is one such effect. By focusing on the yield benefits of fertilizers, scientists neglected to ponder just what became of all that applied nitrogen fertilizer.

The reductionist approach to agricultural science is partly a legacy of the German chemist Justus Liebig.[22] Liebig discovered that certain mineral nutrients are necessary for plant growth, especially nitrogen (N), phosphorus (P), and potassium (K). Liebig found that when nutrients are in short supply, plant growth can be stimulated by increasing the supply of the nutrient that is in shortest supply. In other words, if nitrogen is below the level a plant needs for maximum growth, add nitrogen. Liebig's work in plant nutrition did much to help us understand differences in fertility among different fields. Its limitation is that this information is used to treat symptoms rather than as a diagnostic tool. Rather than addressing why a soil is low in a nutrient and what can be done to get the soil to maintain a more "fertile" nutrient balance, a complex ecological problem, the reductionist approach simply addresses the symptom of unbalanced nutrients and recommends a certain N-P-K formula fertilizer to treat the symptoms. This approach acknowledges only one factor that needs attention rather than a whole system in need of adjustment. It is analogous to taking medications to relieve symptoms rather than addressing the underlying causes of ill health. Obviously, this is the prevailing cultural perspective in the United States today. Most people take a couple of aspirins for a headache rather than attempt to reduce the source of tensions that caused it in the first place. For a long time now, farmers have been following scientists' recommendations and applying fertilizers liberally to relieve their soil's symptoms of fatigue and stress. As noted earlier, this approach has created many problems. Unfortunately, little research has addressed the fundamental question "How can we farm without wearing out and using up the soil?"

Quantification. In the search for simple, elegant portrayals of the rules of nature, scientists have relied on numbers. But nature often does not fit neatly into numbers. In order to get the numbers, many value-laden choices must be made. What to measure? What *not* to measure? When to ignore the immeasurable aspects of nature? Statistical techniques help scientists interpret numerical results "objectively," but statistics have their own hidden value choices. For one thing, the use of statistics requires that the scientist decide what chance to take at being wrong. There are (at least) two ways to be statistically wrong: (1) to say there is an effect when there is really an artifact of methods or chance, and (2) to say there is no effect when there really is one, but the methodology failed to detect it (nature's "noise" interfered, perhaps). Scientists prefer the second. To say "I didn't detect an effect" saves face better than "I was fooled by the numbers." The second tends to place the responsibility for error on technology or other external factors (research funding level, for example), whereas the first tends to suggest poor judgment on the scientist's part. So scientists adjust statistics to allow only a small chance of the first type of error and a larger chance of the second. This bias protects the status quo.[23] It requires that an effect be rather obvious before it is accepted as real. The point here is that if we are too much impressed by statistically significant results, we may forget to examine the larger significance of the original research question and the approach taken to answer it. It is critical to ask whether the approach allows for discovery of subtleties and interactive effects, or whether it is biased in favor of the status quo from the beginning.

To the extent that a reductionist, quantified approach to agriculture has led us to narrow scientific and technical solutions to narrowly defined problems, it has prevented our conception of a permanent agriculture designed to work within the laws of nature and the bounds of limited resources. This approach has produced remarkably higher yields with inorganic fertilizers, hybrid crop varieties, pesticides, and specialized one-product farms, but it has also produced the accompanying array of problems described earlier. These problems are complex and intertwined and unlikely to be solved by traditional take-apart reductionist methods. It will take a new coordinated, holistic approach to find solutions that lend permanence to agriculture.

Conquering Nature. In combination with reductionism and quantification, the belief that people can master nature and manipulate and improve upon nature for their own ends was the impetus for the scientific, industrial, and technological revolution of the past four centuries. Writing early in the seventeenth century, Francis Bacon eloquently supported this view: scientific knowledge, and the mechanical inventions to which it leads, do not "merely exert a gentle guidance over nature's course; they have the power to conquer and subdue her, to shake her to her foundations." Bacon writes: "I am come in very truth leading to you Nature with all her children to bind her to your service and make her your slave." However, the path to conquering nature lay in understanding her laws and obeying them: "For the chains of causes cannot by any force be loosed or broken, nor can nature be commanded except by being obeyed."[24] It seems that the first part of his message was given more heed than the second, for modern science has been fundamentally concerned with fighting nature's laws. This is readily apparent in the agricultural sciences, even in the language used in addressing problems: scientists speak of "battling" weeds, "eradicating" pests, even "combating" erosion.

People could manipulate nature as they desired, without paying consequences, because they were superior, according to Bacon's reasoning. From superiority it was a short but dangerous leap to believing that humans were independent from nature. To forget that humans are but one component of the biosphere, and therefore dependent on nature, is dangerous indeed. This is the belief that allows us to forget what to do with our own wastes and to become so enamored of our technical products that we neglect to protect and save their natural sources. It allows agriculture to embrace inorganic fertilizers and hybrid crops with little thought of the consequences to their ultimate sources—fossil fuels, the soil, and wild species and their habitats.

Objectivity. By viewing humans as above nature, as free of nature's constraints, scientists gain a so-called objective viewpoint. The key to this objectivity is the separation of humans from nature. Belief in this separateness implies that scientists can see nature without self-involvement, without value prejudice, because nature is something other, something outside of self. Objectivity has come to mean trusting only carefully controlled, measurable, quantitative results. Qualitative observations are considered too subjective. This science is supposed to be value free, purely truth-seeking. However, the emphasis on objectivity itself is based on particular values and beliefs, rather than on absolute truths. To accept the objective viewpoint one must clearly *believe* in humans' separateness from nature and *value* that separateness as good, right, and proper.

From the position of superior, objective observer, it is easy to look upon nature without feeling connected or responsible to nature. It is easy to lose sight of the interconnections, mutual dependence, and responsibility to nature as the sources of physical sustenance on the planet. Being a superior, purely objective observer of nature is not compatible with a nurturing role, which admits responsibility and requires self-involvement and is therefore inherently subjective. As men no longer saw themselves as keepers of the garden, as nurturers of nature, but as lords over, as exploiters of nature, nurturing came to be labeled a feminine trait, with no place in objective, masculine science. The lack of nurturing in the science subsequently practiced, and the agricultural system that grew out of it, is an important key to our present problems in agriculture. If we approached agriculture with more of a nurturing attitude, we would create a fundamentally different research program, right down to the questions researchers ask. For instance, pests would no longer be seen as the enemy that must be eliminated, but as an inevitable part of the agroecosystem, albeit a part that must be controlled. High numbers of pests would become the fever, the warning symptom that all is not well within the system. Researchers would seek the causes of ill health rather than seeking simply to eliminate the symptoms. They would no longer ask, "How can we kill this pest?" but rather, "Why is this species' population so high? What happened to its natural enemies? What is it about our crop system that is so attractive to this species? What can we do to bolster the health of the system, improve its resistance to this pest, and reduce the pest's numbers?"

ECONOMIC VALUES AND ASSUMPTIONS

The economic goals of agricultural research come from a reductionist legacy also. The scientific goal of increased yields fits neatly with the economic goal of increased productivity.

Unlimited Growth in Productivity. Agricultural research has steadfastly striven for increased productivity of the agricultural sector by increasing the production of farm products, or reducing the amount of input of some type that is required for a certain amount of product, or adding value to raw farm products in off-farm processing and packaging. The accounting has been sloppy, so that increasing the amount of land a farmer can plow in an hour is considered an increase in productivity, despite the fact that the increase required a very large increase in energy input in the form of fossil fuels for manufacture and operation of farm machines. Often, increased productivity is achieved by substituting one input for another that cannot be measured in the same units. This makes it easy for certain costs to be overlooked.

When agriculture is treated as an industry and judged by industrial economics, efficiency, measured by profits, becomes the standard for survival. Only the most efficient farmers survive. An economics-based value system concludes that this outcome is for the common good, with "good" being measured in terms of dollars generated per capita.[25] This conclusion ignores the uneven distribution of those dollars among those "capita" and certainly ignores that "good" could involve something more than dollars. It could mean peace of mind, contentment, harmony with nature, connections to past and present and people and places, or knowledge and care for particular pieces of land. These nonmonetary factors cannot be given dollar values and therefore do not figure into economic analyses of agriculture, but they do figure into how farming is practiced. If agriculture is purely a science, then farmers are replaceable parts in the industry. But if farming is at least partly an art, a cultural practice, then experienced farmers are part of the agricultural knowledge base and should not be discarded.

Short-term Economics. The belief that the common good is served by individual profits fueled the growth of the agricultural input and postfarm processing sectors. Research supported this growth by taking on the short-term time frame of economics, measuring the benefits of various methods, or the choice of research projects, or policy decisions on the basis of short-term returns.

This emphasis on short-term economic returns is a major obstacle to the adoption and maintenance of good stewardship measures. It is hard for farmers to justify to their bankers the initial cost of soil conservation methods (terracing, for example), because the return, the value of soil saved, is not only difficult to express in dollars and cents, but must be measured over a span of decades. This is the moral dilemma faced daily by Soil Conservation Service workers and by farmers in recent years—soil conservation has not paid in terms of short-run economics.[26] A farmer who has already invested in machinery that makes contouring impractical may not even be able to justify contour plowing economically. Indirect expenses add up too. Grassed waterways take land out of production and reduce total farm yields. Strip-cropping the Palouse in Washington State means a farmer cannot use aerial crop dusting because the pilots simply will not fly strip-cropped fields for weed control.[27]

Similarly, the decision to go with pesticide applications rather than integrated pest management or cultural methods of controls stems largely from overriding concerns for the immediate harvest prospects rather than consideration of long-term side effects, potential clean-up costs, or effects on harvests several years later. Many farmers have had little choice because of high debts and marginal incomes. Often bankers have made their decisions for them. Thus the economic values used to judge success in agriculture are often at odds with good stewardship. And farmers are frequently trapped in the middle, desiring to practice stewardship but caught by the need to comply with short-term financial obligations.

The problem is that the short-term time frame generated by the profit motive is incompatible with the longer time scales inherent in natural biological processes. Most of the ecological problems of agriculture developed over a matter of decades, and most will likely take even longer to reverse. Soil erosion does not show up as reduced productivity from one year to the next. Variation in weather is likely to have much more obvious effects and to mask the slow, insidious effects of erosion. It takes even longer to replace lost soil, on the order of tens of decades. Similarly, it takes time for pesticide resistance to appear, for low-level pesticide exposure to translate into cancers, for irrigated soils to turn salty, or for an aquifer to stop yielding. How can loss of genetic diversity, or eroded soil, or groundwater contamination, or loss of predator populations be evaluated in terms of short-term profits and losses? This dilemma inhibits research into the long-term effects of farming practices. Such research may not provide any returns that can be measured in the short time frame of research grant awards or tenure and promotion schedules. It inhibits thinking in terms of developing farming practices that will sustain the natural resource bases of agriculture and permit the earth to be farmed indefinitely. While short-term economic considerations cannot be ignored, they should not set the bounds for agricultural decisions that have ecological impacts. Chapter 3 explores in more detail the inappropriateness of a short economic time frame to the concept of ecological sustainability.

Industry or Ecosystem? Thus, a predominantly economic interest in agricultural science, along with an approach to science that tends to take the world apart and simplify it to understand it, and a cultural world view that places humans as irresponsible masters over nature, have combined to create an agriculture that resembles an industry more than an ecosystem and that is not sustainable. Yet agriculture is essentially biological. Corn simply cannot be produced without living corn plants. And agriculture is essentially ecological. Corn plants grow in an environment with complex interactions occurring with the living and nonliving media around them. What happens if society changes its perspective and looks at agriculture as an ecological system? What could this reveal about the current crises and about what to do to achieve an agriculture that can remain productive indefinitely?

SHIFTING PHILOSOPHIES: THE ECOSYSTEM PERSPECTIVE

What is an ecological system? An ecological system, or ecosystem, is made up of all the organisms in an area, their environment, and all their interactions. Environment refers to both biological and physical factors. Biological factors include the organisms that compete with, benefit, or otherwise interact with one another. Physical factors include such phenomena as climate, moisture, and characteristics of the mineral soil. Of course, the living and nonliving components of the environment exist in a constant interplay. Inorganic mineral nutrients cycle from the soil through organisms and back again, from the nonliving to the living world, as organisms grow and die. Organisms must respond to physical aspects of their environments but also have great influence on these same factors. Organisms can influence local, regional, and even global climatic factors as well as such other factors as the availability of soil moisture and nutrients.

The word *ecological* implies great complexity, fuzzy boundaries between the biological and the physical world, and interactions that are interwoven and can be understood only by attention to the whole picture. Cycles of life and death, seasonal events, and rare catastrophes produce the rhythm of ecological processes. Often, changes occur very slowly over a long time span. Thus, ecology, the study of ecological systems, requires a "whole picture," or holistic approach, and a long time frame of reference. These features, holism and a long time frame, are critical aspects of the sustainability of natural ecosystems and are central to the concept of an ecological approach to building a sustainable agriculture.

By definition, sustainability refers to the long-term endurance of a system. The essential feature of ecological sustainability is the preservation of nature—that is, the plants and animals, as well as the soil, air, and water on which all organisms depend for sustenance. Any sustainable agriculture must endure indefinitely without depleting its ecological support base. This means that a sustainable agriculture must retain and build soil, and it must be maintained without a dependence on finite fossil fuel supplies and synthetic chemical inputs.

To incorporate the concept of ecological sustainability into agriculture, then, requires a fundamental shift from the economic perspective that has guided scientific agriculture for the past hundred years or more. The ecological perspective differs with respect to both the complexity of factors involved in the system and the long time frame of consideration. Whereas the ecological perspective appreciates the complexity of natural ecosystems, the traditional economic/scientific approach attempts to simplify agroecosystems and ignores many factors entirely. An ecological time frame, in contrast to the short time frame of economics, permits the tracking of slow, insidious losses and beneficial effects that take more than a fiscal year to show up. An holistic, long-term perspective is simply more appropriate to biological systems and essential for gaining an understanding of sustainability in agriculture.

HOLISM

The kinds of questions ecologists ask about natural ecosystems reflect the differences between an ecological perspective and the modern agricultural perspective. Ecologists ask how ecosystems function, how they are sustained by sunlight, how species interact and coexist, and how energy and materials circulate within and between adjacent ecosystems. A move toward a sustainable agriculture implies that similar questions ought to be asked about agroecosystems. Researchers should be attempting to understand the functioning of agroecosystems rather than simply trying to manipulate that functioning. They should be asking how to make agroecosystems function more on sunlight and less on fossil fuels. They should investigate whether some crops might grow better together rather than alone in certain cases. And they should be trying to obtain efficient circulation of energy and materials within agroecosystems and to minimize or take advantage of exports to adjacent ecosystems. These types of questions contrast with the current emphasis on high productivity in industrial agriculture.

Adopting an ecological approach to agriculture means a shift toward dealing with unique circumstances and a simultaneous shift away from general formula solutions applied over a broad range of conditions. Conventional agriculture practitioners have tried to homogenize the physical aspects of environments with nutrient, chemical, and water inputs. Scientists have adopted elaborate assumptions to explain how diverse biological systems in different locations can be treated identically.[28] This approach leaves the full functioning of agriculture's biological machinery unknown. Research in sustainable agriculture would instead stress the uniqueness of time and place and of working with local natural resources and processes. The ecological perspective creates awareness that the complexity and interconnectedness of organisms and environment make complete predictability of system behavior difficult, if not impossible. It reveals the need to allow for variation in time and space when designing sustainable agroecosystems.

According to the holistic perspective, complex systems cannot be understood by taking them apart and then reassembling them. Sustainable agriculture must seek what Wendell Berry in *The Gift of Good Land* refers to as "patterns that work" rather than single cause-and-effect relationships. This runs counter to the instincts of those who set out to manage biological systems, for simple cause-and-effect relationships are easier to manage. Complex biological systems do not easily reveal such simple relationships, nor are they always easily understood. Yet they often work well. A holistic approach involves examining agriculture at the whole farm or ecosystem level. For example, it may include looking at nutrient inputs and outflows, overall farm biological stability, and changes in the soil over time. By looking for overall patterns that work, this approach attempts to incorporate complex natural relationships into agriculture without first taking them apart and looking for cause-and-effect relationships. Rather than perfecting one crop at a time, the holistic ecological perspective suggests seeking out a collection of plants and animals that work well together.

Sustainable systems must be designed at the ecosystem level as well. Whereas the process of domesticating a crop for monoculture occurs by artificial selection at the population level, domestication of a crop that works well within a whole farm agroecosystem must take place at the community level. This means that it must be selected in the context of interactions with other species. This type of selection has been practiced in some diverse agroecosystems maintained for centuries by native cultures in the tropics (see Chapter 4). In these cultures, the crop species have coevolved with one another, with plant-eating insects, weeds, and, to a large extent, with the people who have nurtured and tailored them over time.

The goal of sustainability—in other words, the preservation of the ecological resource base—when applied to agriculture contrasts starkly with the goals of industrial agriculture. Perhaps most obvious is the way soil is regarded under these contrasting goals. If sustainability is the goal, the soil, as the prime determinant of an agroecosystems's carrying capacity (the population size supportable by a given environment), becomes the major focus of research and stewardship. In contrast, when industrial efficiency is the goal, the soil is regarded as merely a medium for conveying applied chemicals to plants and other organisms in order to produce maximum crop yields. A much higher priority would be given to environmental quality, yield stability, and internal control of fertility and pest damage if agriculture adopted the goal of sustainability. A higher value would be placed on maintaining soil if it was recognized as the foundation for a healthy ecosystem (see Chapter 3). Agricultural research must become less concerned with the production of marketable goods, and more concerned with the careful management of the ecological resource base, in order to achieve sustainability.

LONG TIME FRAME

Ecological studies consider longer time scales than those of economics. Many processes and cycles in nature approximate or exceed a typical human life span. Examples are the life spans of some herbaceous perennials,[29] the periodicity of devastating hurricanes in eastern deciduous forests,[30] wet/drought cycles on the prairie,[31] volcanic catastrophes and recovery of the plant communities affected, evolutionary change, and the succession of natural communities to their climax state. As stated earlier, a short-term economic emphasis is unlikely to address the problems of soil loss and pollution inherent in modern farming practices. This is because environmental degradation is monetarily cost-effective, at least for a short while. Again, sustainable farming must adopt a view that preserves the land for many generations rather than exhausting it for some short-term profit. The soil is much older than any living person. Does anyone have the right to destroy in a few generations what may have taken millennia to build?

In summary, it is clear that sustainable agricultural models must incorporate both holistic approaches and long-term perspectives into their formulations. As discussed in Chapter 6, new forms of agriculture will only become compelling when the agricultural establishment decides that environmental and social considerations override some short-term economic goals. What is suggested here, however, is more than simply ecology coming to the rescue of agriculture. Multidisciplinary research agendas are needed. Otherwise, some sustainable agroecosystems, though biologically and agronomically viable, will fail because of the lack of an appropriate cultural context. Hence, farmers, economists, sociologists, and anthropologists must all play a part in finding solutions for the problems of agriculture. The next chapter turns to a discussion of what nature can teach us about sustainability on this planet.

3

An Ecological Perspective on Sustainability

THE PRECEDING CHAPTERS have established that current agricultural practices are biologically and economically unsustainable and that these practices are rooted in the science, research institutions, government policies, private enterprise, economic system, and underlying philosophies and values of society. Given these roots, it is not surprising that conventional institutions have been unable to address adequately the problems of agriculture. The ecological perspective has been introduced as a perspective that is inherently more compatible with a goal of sustainability.

This chapter takes off from that point to explore the ecological aspects of sustainability and to outline the basic changes in perspective that this concept requires of agriculture. Because the failures of industrial agriculture have underlying biological causes, this chapter examines what ecology and related sciences can explain about the sustainability of natural ecosystems. Natural ecosystems can show us how nature maintains a dynamic equilibrium and circumvents the environmental problems now encountered by agriculture. Nutrient dynamics, energy flow, soil building, successional changes, population changes, interactions between species, disturbance, and plant life histories are important aspects of natural dynamics discussed here. The chapter closes with a summary of some important ways in which conventional agriculture differs from natural ecosystems in terms of sustainability.

ECOLOGICAL INSIGHTS INTO SUSTAINABILITY

How natural ecosystems regulate crucial materials so that inputs approximately equal outputs in a steady state is a primary concern in ecosystem studies. If agriculture is to remain productive without destroying its resource base, a steady state must be sought for agricultural ecosystems. An understanding of how the dynamics of natural biological systems are regulated will lay the foundation for building sustainable agricultural systems. In natural ecosystems, production is tuned to available levels of energy and nutrients. Hence, natural ecosystems provide the best models for the structural patterns necessary to achieve the tight nutrient cycling and solar-driven energy flow that are crucial to agricultural sustainability.

All the living organisms in an area plus the physical environment with which they exchange materials and energy constitute an ecosystem. A forest ecosystem consists not only of trees, birds, mammals, insects, and the fungi and microbes crucial for decomposing organic material and recycling nutrients, but also the parent material beneath the soil that provides minerals essential to the system, water that enters as precipitation, and the solar energy captured through photosynthesis and transferred from organism to organism within the community. An ecosystem is considered mature when it has achieved some sort of dynamic equilibrium. In mature ecosystems, minerals cycle around some steady state. In other words, mineral inputs on the average equal exports, and the living portion of the ecosystem is in a dynamic equilibrium. In such balanced ecosystems, overall biomass—the total mass of living organisms—neither increases nor decreases over time.

Chapter 4 develops the theme of nature as a measure for agriculture. However, to use nature as a model or standard for sustainable agriculture, it is first necessary to appreciate the features of natural ecosystems that enable sustainability or permanence. It is necessary to decipher the ecological functioning and patterns of natural terrestrial ecosystems; to become familiar with the structure and dynamics of the communities and populations that comprise ecosystems; and to discover in natural ecosystems the kind of features that must be incorporated into or imitated by agricultural systems to make agriculture more sustainable. This chapter provides a primer in ecology as background before proceeding with Chapter 4's argument for incorporating natural processes into agriculture.

ECOSYSTEM PROCESSES: ENERGY FLOW AND NUTRIENT CYCLES

Ecosystem organization and function depend on the processes of energy and nutrient flow. The ultimate source of the earth's energy is the sun. Solar energy striking the earth is stored in chemical bonds via photosynthesis and then flows through food webs. Nutrients arise from various pools—for example, calcium, phosphorus, and iron from the lithosphere, carbon and nitrogen from the atmosphere, and water from the hydrosphere. Although, in many cases, the flow is one way, these materials recycle many times through the biosphere before they are lost; so resupply rates can be slow.

Organisms can be classified into three broad groups, depending upon how they obtain most of their food. Primary producers are those that transform energy, primarily radiant energy, into chemical energy. All green plants and a few bacteria are primary producers. Consumers are those organisms that derive all of their food from other living organisms. They range from animals to parasitic plants and fungi. Decomposers derive their food from dead organisms. They are nature's recyclers, shuttling nutrients from dead to living organisms and reducing dead organic matter to increasingly smaller pieces until it is once again mineralized.

Energy. Energy and nutrients flow from one organism to another when organisms eat. Thus each organism can be thought of as a link in a chain of energy and nutrient flow—a food chain. Each feeding, or trophic, level is one step further from the sun, the ultimate energy source for most of the earth's life. Categorizing organisms into different trophic levels is somewhat arbitrary, since many animals eat at more than one trophic level. For example, people who eat plants are feeding at the second trophic level; people who eat cattle are feeding at the third trophic level; and those who eat carnivorous fish are feeding at the fourth trophic level. Because most animals feed at several trophic levels, or use a diversity of energy sources, it is more accurate to think of trophic relationships as food webs, or series of interlocking food chains.

Energy always *flows* one way along food chains, but nutrients typically *cycle* many times among trophic levels. Decomposer and primary producer organisms are essential to a self-sustaining ecosystem. Consumers, on the other hand, primarily accelerate rates of energy and nutrient flow. They are opportunists, dependent on producers and decomposers and, at least theoretically, not necessary for an ecosystem to function.

Energy and nutrients passing from species to species within food webs influence both the structure and functioning of ecosystems. Energy enters an ecosystem primarily as solar energy and is converted to chemical bonds in plants and some bacteria during the process of photosynthesis (see below, under "Carbon"). This potential energy, existing in such energy-rich compounds as carbohydrates, fats, and proteins, is available to support both consumers and decomposers. Approximately 90 percent of usable energy is lost at each transfer between trophic levels, and therefore a large number or mass of primary producers is needed to support the upper members of a food web. The trophic structure of ecosystems is sometimes expressed by classifying organisms into trophic levels and estimating biomass. When presented graphically, this structure results in a pyramid formation, with the producers forming the bottom layer.

Although the passage of energy is essentially one way, from sun to eventual heat loss from organisms, mature ecosystems may use energy relatively efficiently, transferring it in chemical bonds among many organisms before it dissipates ultimately as heat. Still, net production declines progressively at successively higher trophic levels. In addition to solar fixation during photosynthesis, energy in the form of chemical bonds moves into ecosystems in migrating organisms and detritus (dead plant and animal tissue). Energy is lost from ecosystems as respiration, during trophic exchanges (feeding), and when exported in moving organisms and detritus. For example, some salt marshes support fewer trophic levels than would be expected, judging from the amount of plant biomass produced, due to their high export of materials with each outgoing tide.

Ecosystem theorists[1] have begun to recognize that, in a sense, some energy is also recycled as it flows through an ecosystem. As mentioned above, at each transfer most of the available energy is lost and the rest is transformed to a condition of higher quality. Hence, along an energy passage, the quantity of available energy decreases, but the quality and biological information associated with the remaining energy increase. Some of this higher-quality energy feeds back to reinforce the processes that enable the system to self-regulate. For example, large animals control smaller organisms through behavior, foraging strategy, waste products, pollination, and seed dispersal. In these ways, energy contained in by-products recycles back to reinforce the ecosystem's productive process.

The earth receives huge amounts of energy from the sun each day. Much of the sun's energy, however, is absorbed by the atmosphere before reaching the earth, and some is reflected back into space. Finally, approximately 1 percent of the sun's energy that reaches the earth's surface is converted into biological energy by living organisms. Yet this small fraction is responsible for driving virtually every biological process on the planet, from the production of high-energy molecules to the construction of carbon building blocks within plants. Animals then consume plants to obtain energy and nutrition, and other animals eat those animals, and so on through the food chain. As noted earlier, during the transfer of energy from one trophic level to another, approximately 90 percent of the food energy is lost as heat. With only this 1:9 efficiency-of-conversion ratio between trophic levels, then, it would take, for example, a hundred kilograms of prairie grass shoots to feed ten kilograms of field mice to support one kilogram of Red-tailed hawk. This conversion ratio explains why top carnivores typically require large hunting areas. It also helps explain why food chains are generally short, with only four or five trophic levels supported.

Although, when ecosystems are at steady state, incoming solar energy sets the upper limit on the amount of living matter supportable, the rate at which plant matter is actually produced in an area is to a large extent dependent upon temperature and rainfall. These climatic factors often limit production to a much lower rate than would be predicted based on available solar radiation. Constraints on plant production include water shortage, shortage of essential nutrients, too high or too low temperatures, and shallow soil.

Ecosystems require constant energy input because of the principles of energy conservation that physicists call the first and second laws of thermodynamics. In short, the first law states that energy entering a system is either stored or translated into work. The rate at which plant matter is produced is called primary productivity and is the primary way an ecosystem stores energy entering the system. Energy is translated into work in the ecosystem when organisms metabolize food to grow, move, reproduce, and maintain their health.

Collectively, this process of losing energy during metabolism is called respiration. When primary productivity equals respiration, then the system just maintains itself, and there is no change in stored energy content and no increase in biomass. When productivity exceeds respiration, biomass can accumulate. The extra productivity is called net primary productivity by ecologists, or yield by agriculturalists. The second law states that at every exchange of energy, some becomes unavailable for work. Therefore, at every exchange between trophic levels, some energy is lost from the ecosystem. The constant input of solar energy keeps the system from winding down.

Photosynthesis is the process whereby incoming solar energy is converted to stored chemical energy. It is a complex process that consists of two sets of reactions, the first of which depends on light, whereas the second depends on the availability of carbon dioxide. The first set, known as the light-dependent reactions, involves changing the sun's energy into biological energy by adding a phosphate molecule to ADP (adenosine diphosphate) to make ATP (adenosine triphosphate). A great deal of energy is stored in this phosphate bond. Molecules of ATP are then stored until the organism needs their energy, when ATP molecules are converted enzymatically back to ADP. In a similar way, high-energy electrons derived from the splitting of water molecules by light are stored for future use in other complex molecules. The second set of reactions are called the carbon-fixing reactions, because they are responsible for bonding carbon atoms to each other and to other atoms to form a variety of carbon-based molecules, which are the chemical storage units and foundation molecules of all plants and animals. The carbon compounds thus produced provide the energy source for almost all living organisms and serve as the basic carbon skeleton for all organisms.

Hence, all life on earth depends on the ability of plants to translate solar energy into edible and nutritious food. Ultimately, agriculture will have to pay attention to the availability of solar radiation, as well as other ecosystem constraints on productivity, as fossil fuel supplies give out.

Nutrients. The flux of nutrients is the second major organizing process in ecosystems. As we have seen, nutrient and energy flows are tightly coupled, as solar energy is transformed into carbon and phosphate bonds and stored in plants and then transferred among organisms along food chains. The movement of energy and matter among trophic levels and between physical and biological components integrates an ecosystem into a functional unit. As the bonds are broken during respiration, the compounds involved are degraded and the constituents are released. By its very nature, energy cannot be recycled. Each unit of energy can be used only once before it is lost as entropy. Life on earth is possible only because of a fresh supply of solar radiation every day. The abundance of life depends on this fresh supply, but also on nature's thriftiness and ability to store and transfer energy conservatively. In contrast, nutrients are not only conserved but recycled in nature. If plants and their consumers were not eventually decomposed and recycled back into the food chain, the supply of nutrients would become exhausted and the ecosystem would cease functioning. Organisms extract nutrients from their environment, hold on to them for a period, and then lose them again.

Of all the elements required by organisms, the most common are oxygen, hydrogen, carbon, nitrogen, phosphorus, and sulfur. Other important elements include potassium, calcium, magnesium, iron, and manganese. Many of these latter nutrients arise from soil's parent material—crushed rocks—and are brought up by plant roots. Three of the most important nutrient cycles, and those with much relevance to agroecosystems, are the nitrogen, carbon, and phosphorus cycles.

Nitrogen. The nitrogen (N) cycle is the mineral cycle most susceptible to change. Although the atmosphere, at 78 percent nitrogen, is by far the most abundant source of N on earth, with nearly 8000 kilograms of N_2 gas above every square meter (or 1640 pounds per square foot) of the earth's surface, nitrogen is available for uptake by plants in only two forms, nitrate (NO_3^-) and ammonium (NH_4^+) ions. Because N_2 gas is very stable, however, it takes a great deal of energy to split the triple bond between its N atoms so that the N can be incorporated into a biologically useful compound. Some atmospheric N enters the ecosystem when it is oxidized by lightning. The oxidized N then reacts with atmospheric water and falls as dissolved NO_3^- in precipitation.

The primary biological way in which N enters the ecosystem from the atmosphere involves the nitrogenase enzyme, which is possessed by a few free-living bacteria, actinomycetes, and cyanobacteria in litter and upper soil layers. The vast majority of nitrogen-fixing microbes, however, are *Rhizobium* bacteria symbiotic with legumes. The nitrogenase enzyme enables these microbes to convert atmospheric

N to biological N compounds directly. *Rhizobium* bacteria infect roots, inducing the formation of nodules where they reduce atmospheric N using carbohydrates provided by the host plant as an energy source. Hence, leguminous plants "pay" carbon for nitrogen. This strategy is advantageous where soil nitrogen is particularly low. A few other nitrogen-fixing symbionts are actinomycetes (*Frankia*, for example) that associate with such woody plants as alder (*Alnus*), buckbrush (*Ceanothus*), or members of the rose (Rosaceae) family. Typically, legumes can account for 100 to 300 kilograms of N per hectare (or 90 to 270 pounds of N per acre) each year, and alders can account for 80 kilograms of N per hectare (72 pounds per acre) each year.

In the soil, ammonium (NH_4^+) is oxidized biologically to nitrite (NO_2^-), then to nitrate (NO_3^-), by certain bacteria called chemoautotrophs, which means that they live off the energy released by the conversion of NH_4^+ to NO_3^-. The simultaneous release of hydrogen ions (H^+) acidifies the soil. Nitrate and ammonium are taken up directly by plant roots and soil organisms and used in the formation of proteins and other nitrogen-containing compounds.

Nitrogen in these organic forms then passes from organism to organism within food webs. When an organism dies, its tissue sloughs off, or when wastes are released, the nitrogen-containing organic mater enters the detritus food web—that is, the community of animals, fungi, and microbes involved in decomposition. There it is eventually mineralized to ammonium ions (NH_4^+), a form once again available to be taken up by living organisms. Hence, the availability of NH_4^+ and NO_3^- ions is determined by the activities of a variety of microbes, including nitrogen-fixers, nitrifiers, denitrifiers, and ammonifiers, whose relative activities are largely dependent upon and vary with such soil factors as percentage of organic matter, moisture, texture, chemistry, and pH. As agricultural practices alter these factors, the activities of the soil community, so crucial for regulating nutrient cycling and availability, are inevitably altered. Understanding how nutrients cycle in natural terrestrial ecosystems and in agricultural ecosystems provides clues for achieving tighter nutrient cycling in agriculture.

Nitrogen is lost from ecosystems in a variety of ways: through leaching, and export of organic matter, denitrification, and ammonia violatilization. Nitrogen leaches, or moves downward through the soil, primarily as NO_3^-. Since NO_3^- is not electrostatically attracted to the negatively charged clay and humus particle surfaces in most soils, it moves freely with soil water and is thus very susceptible to leaching. Once this nitrate descends beyond the effective reach of plant roots, it is lost from the ecosystem. When soil nitrogen levels are excessive, as in fertilized fields and cattle feed lots, excess nitrate can leach into groundwater, where it contaminates water supplies (see Chapter 1). In contrast, when nitrogen contained in organic matter mineralizes to NH_4^+ during decomposition, this NH_4^+ associates with the negative charges on soil particles and is thereby held in the soil.

The export of organic matter also results in a loss of organic nitrogen from ecosystems. Exported materials may consist of plant or animal tissue, or soil organic matter lost through wind or water erosion.

The third means of nitrogen loss from ecosystems is denitrification—the conversion of ionic N to N_2 or N_2O gas—which occurs primarily in low-oxygen (anaerobic) soils. Denitrification is caused by certain anaerobic organisms that use nitrogen instead of oxygen as their final electron acceptor. Rates of denitrification increase with such factors as high NO_3^- concentration, availability of carbon sources, temperature, and low soil oxygen (O_2). Thus, denitrification proceeds rapidly in soils that are compacted or chronically wet, and these soils are in danger of being depleted of nitrogen. This is one reason why agricultural practices that compact the soil cause problems for plant growth.

The fourth means of nitrogen loss from ecosystems involves the volatilization of ammonia (NH_3) gas. Conversion of urea from dead plant or animal material or urine to NH_3 occurs fairly rapidly in the soil. This explains why catttle or bison urine represents a relatively insignificant nitrogen input into grasslands. It is manure, more so than urine, that is the source of nitrate pollution of the groundwater in animal feed lots.

Carbon. The carbon (C) cycle is similar to the nitrogen cycle in that a large carbon pool exists in a gaseous form within the atmosphere in the form of carbon dioxide (CO_2). Carbon dioxide, however, constitutes only about 0.04 percent of the atmosphere. Atmospheric CO_2 is taken up directly by green plants during photosynthesis. Plants split CO_2 and water molecules and rearrange their components to form carbohydrate (glucose) for energetic and structural purposes.

In maturing ecosystems, in which overall growth exceeds decomposition for a time, carbon accumulates in plant and animal biomass and in soil organic matter to some maximum supportable by the ecosystem. In grasslands, most of this accumulated carbon occurs underground as roots, rhizomes (underground stems), and soil organic matter. In many forests, the carbon storage system is inverted: most of the stored carbon is in material standing above the ground and the soil is relatively poor in organic carbon. The extremes are seen in peat bogs, where the rooting substrate is entirely organic peat, and in wet tropical forests, where virtually all organic carbon exists in living organisms and is quickly recycled by fungi and microbes due to the warm, moist climate that favors decomposition. Peat bogs represent a substantial global carbon reserve. When peat soils are broken for agricultural purposes, organic matter is rapidly oxidized and soil carbon is lost to the atmosphere. It may seem counterintuitive that tropical rain-forest soils are among the poorest in the world. The lush growth of these forests would seem to imply fertile soil. In reality these soils are virtually devoid of nonliving organic matter, and even root mass is generally low. It is therefore not surprising that as much as 60 percent of the trees in a tropical forest may be nitrogen-fixing legumes. Most roots of tropical plants are finely netted surface feeders that acquire nutrients rapidly as they become available. Some even grow upward to snatch the rain-borne nutrients that trickle along tree trunks before they enter the soil.

Carbon enters the food web of a community as CO_2 during photosynthesis. It may then become a constituent of a sugar, fat, protein, or cellulose molecule. Organisms in the next trophic level incorporate some carbon from the plants they eat, but much ends up back in the atmosphere, dissipated as CO_2 during respiration. Excluding water, about 95 percent of living matter consists of carbon compounds.

Several variations on the process of photosynthesis occur in nature. These variations of photosynthesis have implications for plant community structure and ecosystem function and relate to certain ecological factors. It is helpful, therefore, to review the process and role of photosynthesis in the ecosystem.

Photosynthesis, or synthesis powered by light, is the process by which the sun's energy is converted into biological energy. In its simplest form, photosynthesis can be expressed (with the arrow representing light) as:

$$CO_2 + H_2O \rightarrow CH_2O + O_2$$

That is, carbon dioxide plus water, in the presence of light, yields carbohydrates and oxygen. The oxygen is released into the atmosphere, where it is subsequently consumed for the metabolism of most organisms, including the metabolism within plant cells. Other synthetic reactions further manipulate the basic carbohydrate molecule to create carbon-based fuel molecules, or carbon building blocks.

Hence, carbon dioxide is essential for photosynthesis. It enters the leaf through tiny pores called stomata, which open and close in response to the plant's requirement for CO_2. The stomata are also important in the regulation of the water balance of plants, because water vapor can be lost through them when they are open. On dry days, or during the driest part of a day, the stomata might be forced to close to prevent the loss of too much water vapor. This also has the effect of limiting the amount of CO_2 that can be taken in and hence the rate of photosynthesis.

Three distinct pathways by which CO_2 enters the photosynthetic process have evolved in terrestrial plants. One of the paths begins with the binding of CO_2 to an enzyme to form a three-carbon compound. This is known as the three-carbon, or C_3, pathway, and its rate relies on the rate of diffusion of CO_2 into the plant tissues. Another pathway begins with the binding of CO_2 to an acid to form a temporary compound of four carbon atoms, resulting in a four-carbon, or C_4, pathway. Plants that use this pathway incorporate a physiologically expensive mechanism to capture CO_2, and so C_4 photosynthesis is more energy-consuming than the C_3 pathway. These two pathways, C_3 and C_4, have a significant impact on the ecological characteristics of plants, as we shall see below. The third pathway of photosynthesis is called crassulacean acid metabolism, or CAM, and occurs primarily in succulent plants, such as cacti, of arid ecosystems. Because of their extreme water-conserving strategies, CAM plants display very low CO_2 fixation rates, but their water use efficiencies (water expended per carbon gained) are very high. This is because CAM plants open their stomata at night to gain CO_2, then close them during most of the day, thereby avoiding water stress. At night, the captured CO_2 is stored within the plant's cells as crassulacean acid. During the day, chemical reactions release CO_2 from the acid and feed it into the light-dependent reactions.

Plants that use the C_4 route produce an enzyme that has a greater affinity for CO_2 than does the enzyme involved with C_3 metabolism. Thus, plants that are water stressed, and therefore must keep their stomata closed to prevent further water loss, will benefit from the C_4 pathway, which can most efficiently use what carbon dioxide is allowed in. Conversely, when CO_2 uptake is not limited by the need for water conservation, C_3 plants are significantly more efficient at photosynthesis than are C_4 plants. Temperature affects the propensity for one carbon pathway or the other. Because high temperatures often cause water stress, many plants in hot regions have evolved the C_4 pathway.

Given the conflict between obtaining enough carbon dioxide and conserving moisture, one could predict that C_3 plants would predominate in relatively cool and moist areas or seasons (where open stomata would take in more CO_2 but not contribute to water stress), whereas C_4 plants would be more successful in warm and dry areas or seasons. It turns out that the dryness of a region is a good predictor of the ratio of C_3 plants to C_4 plants. In the extreme, C_4 plants can flourish under conditions that would be lethal to most C_3 plants. In Kansas tallgrass prairie, C_3 and C_4 grasses show distinct seasonal growth patterns. The cool-season C_3 grasses are the first to green up in the spring and turn brown in the summer. The warm-season C_4 grasses, such as big bluestem, stay green all summer and flower in the height of summer's heat. Because the summers typically are hot with a long dry stretch in midsummer, it is not surprising that C_4 grasses dominate then. The percentage of grass species with the C_4 pathway increases from north to south in North America, ranging from 0 percent in Alaska and northern Canada to 70 percent to 80 percent in northern Mexico, Texas, and Florida.[2]

Phosphorus. The phosphorus (P) cycle is fairly typical of many minerals that arise from a parent material (crushed rocks that supply the mineral portion of the upper soil layers). There is increasing recognition by plant scientists of the importance of mycorrhizal fungi in accumulating soil phosphate and making it available to plants via the direct connections between fungal filaments (hyphae) and roots. Mycorrhizae are associations, usually symbiotic, between certain soil fungi and the roots of many plants. They can vastly increase the effective surface area of roots and provide a direct connection between decomposing material and plant roots. They may also facilitate the transport of nutrients directly from plant to plant, even between members of different species.[3] Obviously, learning to enhance mycorrhizal associations in agroecosystems could have major benefits for plant nutrition and nutrient cycling.

Plant roots take up phosphorus in only one form, phosphate (PO_4^-) dissolved in soil water. Thus, the concentration of biologically available phosphorus may be considerably overestimated by conventional soil tests, which include elemental and bound phosphorus. As with many elements arising ultimately from parent material, phosphorus is steadily mined from any ecosystem where export of a nutrient or resource exceeds its input. It is thus an open cycle. Phosphorus eventually ends up as sediment on ocean bottoms, unavailable to terrestrial life. The nutrient loop could conceivably remain open until a geologic uplift raises the ocean bottom and creates a new land formation. Therefore, in a sustainable agriculture, efficient use and recycling of phosphorus can be a goal, but farmers may still find it necessary to replace harvested phosphorus from other sources after a while.

When an ecological community is disturbed, the rate of nutrient cycling may speed up. For example, when a tree falls in the forest, more light and precipitation reach the soil surface around the fallen tree. The uprooting process mixes soil layers. In rapid succession, the rates of decomposition, mineralization, and nutrient uptake by plants increase. This explains the temporary pulse in growth, flowering, and seed production by plants in the understory following tree falls. Fire in grasslands sets in motion similar chain reactions. The fire removes accumulated dead plant material (called litter) and permits more light and warmth to reach the soil surface. This condition promotes growth and nutrient uptake of new shoots in spring. Warmer soil favors decomposer organisms, which leads to some increased mineralization of soil organic matter. Fire is a disturbance that actually is required to maintain the tallgrass prairie ecosystem. On the other hand, fire in many forested ecosystems can lead to the death of dominant trees, release of nutrients in ash, and great risk of nutrient loss via erosion and leaching. As will be explored later in this chapter, periodic localized disturbances can allow more species to coexist, or, in ecological terminology, can increase species diversity. Communities with such disturbances exist as mosaics of sites at different successional stages.

Of course, nutrient cycling is never perfect, and some nutrients are lost by run-off into streams or directly into the atmosphere. Nutrient budgets may take one of three forms. If the budget is balanced, then input equals output, and nutrient status remains at a steady state. Where inputs exceed outputs for a time, nutrients may accumulate in living biomass and dead organic matter. In greatly disturbed ecosystems, such as plowed agroecosystems or clear-cut forests, uptake by plants cannot compensate for erosion and leaching. These are situations where outputs exceed inputs, and the result is a net loss of nutrients from the ecosystem.

Thus, energy is not the only important factor in ecosystem dynamics. Some organisms may be crucial to the flow of a limiting nutrient (for example, nitrogen or phosphorus), even though their role in energy flow is relatively inconsequential.

ECOSYSTEM ORGANIZATION AND FUNCTION

Sustainable ecosystem function depends on regular inputs of nutrients, outputs that do not exceed inputs, and retention of nutrients by the living organisms (the biota) and soil organic matter. These conditions assure that the system does not exhaust its resource base. If nutrient inputs are low, as is often the case in soil-based ecosystems, the retention and recycling of nutrients within the ecosystem is the key to sustainable function. Like the great-grandmother who patches the patches on washcloths, nature is thrifty, even stingy, with nutrients, using and reusing each unit again and again.

Soil. Soil is composed of both physical and biological components. The physical components include mineral particles of various sizes, water, air, and inorganic nutrients. The biological components include soil organic matter and an interacting community of microbes, fungi, plants, and animals all crucial to mineral cycling and ecosystem functioning. A healthy soil has a coarse enough texture to provide air pockets where gas exchange can occur, for roots and most soil organisms require oxygen. The soil must also be porous enough so that water and dissolved nutrients can move through it easily. In compacted soils, gases and dissolved nutrients cannot move easily and soil stagnates.

Soil organic matter is a rich storehouse for critical nutrients. It holds minerals as ions and is the source of mineralizable nitrogen. Thus the fertility of the soil is clearly dependent on the organic matter content. Organic matter also affects the physical characteristics of the soil and protects the soil from damage. Tilth improves, and water and nutrient run-off and erosion decline, when the amount of organic matter in the soil improves. Finally, organic matter is the food source for the decomposer community. All these critical functions occur very near the soil surface in a thin layer where organic matter concentrates. This top layer of the soil, therefore, is where much of the soil's biological activity takes place.

Much of the soil is essentially a colloid, a suspension of clay and humus particles in water. Clay particles are tiny mineral particles, measuring only 0.002 millimeter or less in diameter. In temperate zones clays form from silicates; in tropical soils clays typically consist of iron (Fe) and aluminum (Al) particles. Other soil particles are silt, which ranges in diameter from 0.002 to 0.05 millimeters, and sand, which ranges in size from 0.05 to 2 millimeters. Humus is highly degraded organic matter with particles the same size as clay. Both clay and humus particles have negatively charged surfaces, which bind water molecules and such important cations (positively charged ions) as ammonium (NH_4^+), calcium (Ca^+), magnesium (Mg^{++}), and potassium (K^+). The negative charge associated with humus is generally stronger than that of clay, so humus is the better storage system for mineral nutrients—hence the great importance of organic matter to soil fertility.

Soil scientists typically characterize soil by three layers—the A, B, and C horizons—that correspond to fairly distinct structural and chemical strata. The A horizon is the surface layer. This is the liveliest layer, containing most of the organic matter, soil biota, and low amounts of clay, iron, and aluminum particles. The B horizon, or subsoil, tends to be dense and full of clay. It is a mixture of weathered material and abundant iron and aluminum oxides and clays. In many soils, roots cannot grow well in the B horizon due to the high clay content, high density, low pH, and high aluminum levels. The C horizon consists of a mineral layer that has been affected relatively little by biological soil-forming processes.

Scientists know more about the physical properties of soil than about the biological soil community. Despite a century of ecological inquiry into nature, and the development of explanations of the functioning of natural communities, the biological soil community largely remains a mystery. This is due partly to neglect and partly to the difficulty of studying the subject. Soil communities are not suited to the standard scientific technique of removing components and naming and studying each species separately. The biological and chemical properties of soil depend upon temperature, mineral concentration, and pH, and all these change the moment scientists try to observe them. In fact, because many soil organisms have yet to be described by taxonomists, scientists do not even know what species are down there.

A healthy soil is the basis for a healthy ecosystem. Most of the basic links in food webs are in the soil. The interactions between soil organisms are fundamental to the working of all terrestrial systems; therefore a complete understanding of trophic pyramids should begin and end with the soil community.[4] Herbivore populations are sustained by plants, which in turn are supported by the soil biota, particularly decomposers, mycorrhizae, and nitrogen-fixing bacteria. Mycorrhizae, by increasing uptake and translocation of immobile soil nutrients (phosphorus, for example) to plants, are important in speeding up the cycling of essential elements and minimizing losses by leaching. They potentially exert a major influence on ecosystem dynamics by linking carbon flow and nutrient cycling over a short time scale. Eventually, all biomass not converted by respiration into energy and CO_2 enters the decomposer food chain. In food webs based on decomposing organic matter, such as streams, the bacteria, fungi, and protozoa form the essential link between plant matter and the rest of the food web.

The soil is an ecosystem unto itself. It contains herbivores, predators, and decomposers, and all the accompanying types of interactions among organisms. Symbioses between plant roots and bacteria or fungi occur within the narrow zone of intense biological activity that surrounds roots called the rhizosphere. Parasitism by fungi and herbivory by nematodes and insect larvae on plant roots all take place in the soil. Predators in the soil eat herbivores and have the potential to provide underground pest control. Of all soil organisms, the most familiar are earthworms. These invertebrates benefit soil in a number of ways. Their activities increase aeration, improve water distribution, and enhance a soil's penetrability to plant roots. They are among the first-order decomposers in breaking down litter to smaller organic matter fractions.

It is obvious that soil has great influence upon plants, but over time plants also influence the soil they occupy. Perennial plant roots build organic matter in forest and prairie soils. Much like the annual supply of leaf litter in deciduous forests, sloughed roots of perennials provide a major source of soil organic matter. Acidic litter produced by conifer forests eventually lowers soil pH. These acid soils are prone to leaching of mineral ions and podzolization (formation of a hardpan or impermeable layer in the B horizon). Long-term modification of soil by vegetation means that in some ecosystems, soils may end up suitable for the climax[5] species only.

Healthy soils are alive and dynamic, mixed by earthworms, turned over by burrowing mammals and tree falls, fed by and feeding above-ground communities. No terrestrial ecosystem can long endure apart from an active, healthy soil. But healthy soil communities arise only in minimally disturbed ecosystems. Frequent tillage, as is common in agricultural ecosystems, repeatedly turns back the ecological clock and thus prevents the development of a diverse and functional soil community.

Ecosystem Changes with Time. To mimic nature's endurance in our agriculture, it is necessary to understand how ecosystems change with time, to examine changes in characteristics and processes as ecosystems progress from youth to maturity. Ecologists refer to the various stages of ecosystem development as seres, and the process of maturing as succession. Imagine the processes occurring when the biota restore an area devastated by a catastrophic event such as a hurricane, volcano, or glacier. As the producer and consumer components are removed, the sudden availability of newly dead material is a boon for the decomposer community. Organic matter is mineralized, releasing nitrogen ions and other nutrients into the soil. During the early stages of succession, the area is colonized by weeds and fast-growing, short-lived perennials. These species may quickly cover the soil, putting down roots and maximizing production on the initial pulse of available nutrients. These species, however, eventually give way to seres of slower-growing, but more competitive and longer-lived plants. Eventually the site is dominated by the climax community of plants and animals, which will remain until the next cataclysmic disturbance.[6]

During early succession, then, ecosystems are far from the equilibrium state of mature ecosystems. Nutrients are easily lost by leaching and erosion until perennial plant cover is restored. The potential for soil erosion, and nutrient loss, is great. Overall biomass and species diversity are low. As the system matures, species occupying more specialized niches, or roles, begin to establish. These specialists tend to complement one another in the food they eat or the homes they choose, their light and moisture requirements, or their timing of reproduction and growth. These complementary sets of specialists are more efficient at capturing and using resources within the ecosystem. They fill the available space and use and reuse the available food more fully than the initial colonists. The recycling of nutrients among trophic levels gets tighter, until nutrient cycles become closed. Species diversity, both above and below ground, stabilizes. Efficient resource capture, internal recycling, and a high degree of interdependence among species allow mature ecosystems to support more trophic levels than do immature ecosystems.

As a community approaches its climax, it generally becomes more functionally stable. Stability, as used here, means the ability of the system to return to its equilibrium state after a temporary disturbance.[7] For example, most forests are eventually able to recover from tree falls after big storms, provided the soil and understory community are intact. Usually, sustainable disturbances involve small changes or small areas. The concept of stability is related to resilience, a measure of the ability of the system to absorb changes and still persist as a viable community. However, some climax ecosystems, such as arctic tundra, sand dunes, and tropical rain forests, are among the most fragile in the face of human destruction. This is because the physical and biological components of these ecosystems are in delicate balance. Usually, disturbance leads to loss of soil, nutrients, and species. Any natural ecosystem can be pushed beyond the ability of organisms to restore it. Plants may be unable to restore ecosystem function if permafrost levels change in a tundra ecosystem, or dunes are destabilized in a sand prairie or shoreline, or soil is lost from a moist tropical forest. In a tropical forest, relationships among species are so specialized that many species are dependent on another single species, without alternatives should that other species disappear. The loss of one species can cause the loss of others dependent upon it. Thus, this complex system is vulnerable to disturbance. Complexity alone does not confer stability.

Ecosystem Health and Stress. What are the characteristics of a healthy, balanced ecosystem?[8] In general, healthy ecosystems tend to display high levels of integrity and sustainability. First of all, as stated above, the healthy ecosystem functions around some sort of steady state in biomass dynamics or nutrient flux. Secondly, the complementarity of organisms can confer stability and resilience to the community. Species that occupy complementary niches can compensate for loss of neighbors, hold soil, and ensure that some species will thrive despite variable conditions. Hence, a healthy, diverse community can "track" the disturbance regime or climatic change. In summary, a good *functional* indicator of integrity in an ecosystem is efficiency in energy transfer and nutrient cycling. A *structural* key to ecosystem integrity is a diverse species assemblage in which the longer-lived and larger life forms are dominant in mature phases of ecosystem development.

What indicates ill health in stressed ecosystems? First, primary productivity often drops. Of course, a low productivity value can only be interpreted in the context of the usual values for that ecosystem. Ecosystems differ in their range of normal values for primary productivity, from desert scrub (10 to 250 gm/m^2 per year; equivalent to 90 to 2230 lb/acre per year) to tropical rain forest (1000 to 3500 gm/m^2 per year; equivalent to 8900 to 31,200 lb/acre per year). Still, primary productivity of both these ecosystems typically declines under a stress such as air pollution. Second, damaged ecosystems leak nutrients. Stress often decreases the efficiency with which nutrients are recycled, resulting in losses in nutrient capital.[9] In agroecosystems, deteriorating health is marked by a need for fertilizer to compensate for lost nutrients. Conceivably, this sort of impaired managed ecosystem can remain economically viable over the short run while running down in natural capital. Deterioration of an ecosystem can be temporarily masked by nutrient and energy subsidies, or by artificially restocking sensitive species, but for the system to become self-sustaining, the natural capital must be restored.

A third symptom of ecosystem stress is the loss of certain sensitive species. The first to go are those most sensitive to stress. Thus, stress often leads to the decline of some native species. Although opportunist species may increase, this increase seldom compensates for those lost. Overall species diversity, as well as the proportion of native species within the community, declines. Forests subjected to air pollution show just this response.[10] Similarly, aquatic ecosystems stressed by sewage show increased productivity, but lower water quality and loss of valued fish species. At the same time that sensitive species decline, the population size of various species becomes unstable. Although natural populations may fluctuate as part of their natural dynamic scheme, ecosystem stress can lead to wider fluctuations and potentially uncontrolled instability in population size. Wide swings in population size of a species can lead to extinction of that species.

As discussed in the next section, healthy communities, by virtue of high species diversity, are rarely devastated by disease and insect epidemics caused by resident organisms. Over evolutionary time, native plants adapt to the pathogens and herbivores of a community. In natural systems, severe epidemics are rarely reported except where alien pathogens have been introduced.[11] Any factor that can disrupt healthy ecosystem function, however, can lead to a fourth symptom of ecosystem distress: increased prevalence of insect pests and diseases. Stress weakens the resistance of host species and makes them more susceptible to attack. One quick check on ecosystem health is to measure pest prevalence on vulnerable species. High levels of pests indicate system stress. Air pollution, for example, can encourage outbreaks of forest insect pests. Also, low species diversity, as is characteristic of agricultural monoculture, favors insect and disease devastation.

Fifth, stress can favor smaller, shorter-lived life forms over larger organisms in the community. Ecosystem integrity depends on a biotic structure that features larger organisms. For example, trees maintain forest integrity by modifying the local climate and providing a three-dimensional habitat for the coexistence and protection of many animal species. Under stress, such large forms are often the first to succumb.

Sixth, and last, the presence of contaminants in an ecosystem can be both a cause and a symptom of stress. Human agricultural and industrial activities have led to increased circulation of contaminants in many ecosystems. The health of sensitive species weakens when acid rain or human-made compounds containing heavy metals, insecticides, or herbicides enter an ecosystem. Often these contaminants enter at the bottom of the food chain, but are concentrated at each higher trophic level up the food chain. This bioaccumulation, as we began to learn with DDT in the 1960s, is an indicator of poor ecosystem health.

The health of an ecosystem may be judged by the factors above and by its ability to bounce back after a period of stress. A variety of criteria must be applied to assess ecosystem health, however. For example, a managed forest may provide a sustainable level of timber production, but it may be unable to provide habitats for diverse fauna. This may leave it open to disease outbreaks or nutrient losses. Thus we learn from nature that the complexity common in maturing ecosystems helps tighten the interdependence of organisms, which in turn tightens the recycling of energy and nutrients. But at the extremes, complexity may result in fragility, especially in the face of human-caused change. The natural tendency is to change toward a stable community that features a diversity of long-lived organisms that exist in accordance with the availability of resources. Short, localized, or infrequent disturbances may occur, but these do not disrupt overall homeostasis in a healthy ecosystem.

POPULATION AND COMMUNITY ORGANIZATION AND FUNCTION

Just as nature provides insight into sustainable ecosystem structure and dynamics, it can also illuminate sustainability of other natural groupings. In agriculture, it is especially important to understand the dynamics of the plant components of ecosystems, plant communities, and their components, plant populations. A community is an assemblage of interacting species whose populations are not static, but are continually evolving in response to one another. A population is an assemblage of individuals of just one species. What does nature reveal about the sustainable structure and dynamics of plant communities and populations?

In the interest of simplicity, plants of most terrestrial communities can be grouped into two general categories: those characteristic of early succession and disturbed habitats and those adapted to mature, climax stages. Plants characteristic of immature terrestrial ecosystems are often weedy. They have evolved to exploit pulses of nutrients that arise following disturbance. Natural disturbances tend to be unpredictable in space and time, and so the future favorability of a habitat is uncertain for disturbance-dependent species. Weedy, opportunistic plants have evolved to grow rapidly, put all available energy into seed production, and then die. Their seeds, which may be very numerous and widely dispersed, may then lie dormant in the soil for decades, awaiting the next disturbance. Such plants are the first to colonize fresh disturbances. They are often poor competitors when nutrient levels are low and give way to the more efficient, but slower-growing plants of later seres. Many of these later successional species delay reproduction until they are well established, a period that may take many years. For example, most trees delay flowering until they are tall enough to ensure adequate exposure to light in the canopy. Similarly, many prairie plants emphasize root growth during the first few years to ensure access to deep-soil moisture and nutrients for the long haul, and delay flowering for several years after establishment.

The climax community usually has a greater biomass of living plants than earlier successional stages, near maximum species diversity, populations that fluctuate around steady states, overall balance of gross photosynthesis and total respiration, and recycling of nutrients.

Size and Age Distributions. At equilibrium, populations of wild perennial plants assume a relatively constant array of sizes or ages of individuals, a phenomenon referred to as a stable age distribution. Typically, the array includes a few small, young plants, a few aging senescent individuals, and most of intermediate age. New plants appear just about as frequently as other plants die, so that the average population density (individuals or biomass per unit area) fluctuates around the size of population supportable by its environment, the carrying capacity.[12]

Various factors can influence population size. Some of these factors vary in intensity depending on the population density. These density-dependent factors tend to stabilize population size near the carrying capacity. They work as negative feedback. As a growing population approaches its carrying capacity, population growth slows as each individual's likelihood of mortality increases and probability of leaving offspring decreases. Examples of density-dependent regulatory factors include disease, food supply, and availability of nesting territories. For example, a population of deer with a limited food supply will be reduced as population exceeds the ability of the land to support it. Starvation leads to both higher mortality and lower reproductive rates, and population declines to a density supportable by the environment.

By definition, density-independent factors do not vary with population size and thus do not stabilize populations at carrying capacity. Examples of density-independent factors include such catastrophes as storms, forest fires, and volcanic eruptions. A volcanic eruption is likely to wipe out an entire population in its path, regardless of that population's size or density.

Genetic Diversity. Wild plants rarely occur in nature as dense concentrations of genetically uniform individuals. Rather, natural populations of one species typically are intermixed with populations of other species. In contrast, in modern agriculture, crops are usually grown as single species, or as close to that as the farmer can come. Moreover, natural populations typically are genetically diverse. Genetic diversity, or polymorphism, helps to maintain a stable population size even in the face of diverse selective forces. It keeps a variety of individuals in reserve for the variety of conditions that nature can supply. If drought is severe one year, some individuals will fail; but others, with different genetic make-up, may survive to reproduce, and the population will continue. For example, genetic variation in the length of seed dormancy among individuals allows for the population to survive drastic climatic variability. Even if all the plants germinating in one season die without reproducing because of bad weather, floods, or other disasters, if some seeds remain dormant in the soil and do not germinate, the species may appear again in the same spot the next year, perhaps with more success.

Genetic variability in disease susceptibility is particularly important in wild populations. Three primary determinants of plant disease are the genetic structure of the host population, the genetic structure of the pathogen population, and favorability of the environment for the growth of both. The environment, by influencing the infection potential and growth rate of the pathogen, regulates the severity of the disease and the extent of loss. But a population with genetic variability in disease susceptibility has a buffer against epidemics. This buffering effect works because the pathogen cannot infect every individual, so that the density of susceptible plants is lower than that of the plant population in general. Nonsusceptible individuals physically interfere with effective pathogen transmission, since disease inoculum deposited on them is in essence wasted. In other words, the harder a susceptible host is to find, the less likely a pathogen is to devastate a population. Dense agricultural monocultures of genetically nearly uniform plants make it easy for a pathogen to spread rapidly through a whole field.

In natural communities, many plant species show clumped or patchy patterns of distribution. The patches may be nearly all one species, with individuals packed closely together, or they may be looser, with other species intermixed. Patches may vary as to size and distance apart. This patchiness can have great implications for a population's susceptibility to disease. As patch size increases, it apparently becomes easier for a pathogen to encounter a patch and spread throughout it, so the frequency of permanently infected plants rises.[13] Patch size and arrangement, then, are important design considerations for a sustainable agriculture.

Some plants occur naturally as clones—that is, patches of shoots that all originated from the same seed. Clonal species would appear to be particularly vulnerable to disease because their patches lack genetic variability. How are clonal species able to persist in the presence of disease? Obviously, they survive disease by means other than genetic diversity. Matthew Parker has studied disease dynamics in mayapple (*Podophyllum peltatum*), an herb of deciduous forest floors that forms large clonal populations, or colonies, of limited genetic diversity. This plant is attacked by a potentially damaging fungal pathogen called *Puccinia podophylli*. Despite the presence of this fungal pathogen in most mayapple colonies, the disease has little impact because of an effective physical defense. A whorl of bud scales covers each emerging mayapple shoot and prevents contact between pathogen spores and photosynthetic tissues. This defense permits populations of this clonal species to coexist with its pathogen in nature.

Many weedy plants are escapists. They avoid disease by not persisting in the habitat and thereby failing to provide a long-lived, predictable host. Farmers exploit this natural defense mechanism when they use crop rotations to reduce pest problems.

The Plant Community as a Shifting Mosaic. Mature plant communities are mosaics of patches; each patch has a different history of disturbances and thus differing soil properties, moisture, plant species composition, and microclimate.[14] Because natural disturbances vary in size, type, and frequency, it might be expected that they would affect a community in different ways. For example, it takes longer for natural communities to recover from large or frequent disturbances than from local or rare disturbances. A fire through a forest understory may leave the dominant trees unaffected, whereas a windstorm through a forest may topple the dominants but leave the understory intact. Of course, a forest with trees down will require the longer recovery period.

Because different species are adapted to different successional stages, light conditions, climates, etc., overall community diversity may be increased by periodic, localized disturbances. This periodic disturbance may prevent the loss of weaker species to more powerful competitors.[15] Similarly, grazing mammals that preferentially feed on dominant plants may enable subdominant species to remain in the community, thereby increasing overall plant species diversity. Forest management used to be guided by the principle of eliminating natural disturbances, such as fires. Now foresters recognize that such regular disturbances can serve as productive pulses that are absolutely essential to maintaining a forest's health and vigor. Thus, a community is thought of not as a homogeneous whole, but as a mosaic of differing vegetation mixes. In fact, a community is regarded as dynamic and therefore somewhat difficult to characterize. If an attempt were made to represent a plant community on film, a motion picture would be a more appropriate medium than a snapshot.

Niches. Within natural plant communities, different species typically occupy different niches. In other words, they play particular roles that tend to complement one another. For example, some plants, such as the spring forbs of woodlands, are ephemerals that flower and set seed before the more competitive dominants break bud. In deserts, some plants produce shallow, extensive root systems that capture precipitation soon after it enters the soil, whereas others (mesquite, for example) grow deep storage taproots that enable growth despite drought. Legumes, as well as some other types of plants, form symbiotic associations with nitrogen-fixing bacteria, which provide a source of nitrogen fertility in addition to the soil. Thus legumes occupy a different niche with respect to nitrogen than their neighbors. Many plants rely upon mycorrhizae to accumulate phosphorus and other nutrients, whereas their neighbors may fail to form mycorrhizal associations. In tallgrass prairies, C_3 and C_4 grasses often complement each other where they coexist, with the C_3 using moisture primarily in spring and early autumn and the C_4 actively growing from May through August.

Organisms interact in a variety of ways within communities, sometimes playing more than one role. For example, an adult insect may be an important pollinator of one plant species, but its larvae may be devastating herbivores on another. Or a bird that eats prodigious numbers of damaging insects from a plant in spring may eat that plant's seeds later in the fall. The types of direct interactions most important to plants are competition, herbivory, parasitism, pollination, and symbiosis. Indirectly, plants can benefit from insects that prey upon herbivores, neighbors that provide their own nitrogen, or grazing animals that selectively remove competing plants.

In sum, natural communities tend toward filled niches. These niches, moreover, tend to complement one another.

Competition. Competition arises when two or more organisms are drawing from a common resource pool that is inadequate to support both fully. Plants compete for light by shading one another, for water and nutrients by maximizing root growth, and for space by inhibiting or eliminating neighbors via allelopathy (the direct or indirect harmful effect that one plant has on another through the production of chemical compounds that escape into the environment). Because neighboring plants of the same species have similar forms, shapes, and niches, competition between members of the same species is potentially more intense than competition occurring between members of different species. In present-day communities, rates of competition between different species may be less than expected because species have diverged over evolutionary time to minimize overlap in resource use. Levels of direct competition can be expected to decline overall with ecosystem maturity. This is because colonizing species are adapted to maximize nutrient uptake, growth, and reproduction. They typically do not stay around long after the community recovers from the disturbance. They disappear, dispersing to other disturbed sites or awaiting the next disturbance as seeds in the soil. Climax species, on the other hand, with long histories of evolution in interaction with one another, may have experienced selection for divergence in niche use. Thus, according to this scheme, selection would have led to less overlap of niches among species in mature communities.[16]

Facilitation. Reflecting some prevailing sociocultural values, plant ecologists tend to focus on the negative interactions between plants, especially competition. But plants can also benefit one another. One example, as noted above, is the additional soil nitrogen that can become available to plants growing near legumes. Another example is that of nurse plants, which provide protection for the establishment of seedlings of other species. Nurse plants have been identified in both desert and forest ecosystems. Plants may also benefit from neighbors that attract crucial pollinators, seed dispersers, and beneficial predators of insects. For example, nectar-producing plants on field borders can sustain populations of predatory wasps and syrphid flies, which are important regulators of pest larvae in agricultural fields.

Insects and Diseases. Over evolutionary time, plants have developed the means of coping with organisms that would eat or infect them. Plants employ defensive structures, chemical compounds, or phenologies (seasonal life-cycle timing) to minimize the intensity of the interaction. Indeed, all plants within natural ecosystems are subject to some level of attack; yet populations are rarely driven to extinction. Only when the environment is radically altered by the introduction of an exotic pest for which no defense has evolved, or by the elimination of such natural controlling agents as predatory insects, do plants become endangered.

Plants have four general methods of reducing herbivory: chemical defenses, physical defenses, seed dispersal, and by appearing unpredictably in space or time.

Some plants produce special chemicals that protect against herbivory. These defensive chemicals can work in one of three general modes. Some work as deterrents. Their taste or smell signals insects not to eat the plant. Some work as toxins. Alkaloids are an example of this type. These chemicals cause sickness or death when eaten, even in small amounts. Other protective chemicals act to reduce the digestibility of the plant. Tannins are a good example of this type of defense. They prevent food from being absorbed in the gut, resulting in a slower growth rate for the herbivore. When the herbivore's growth rate slows, it prolongs its vulnerable immature stages. It takes longer for a caterpillar feeding on a rich tannin source such as oak leaves to mature, for example, than one feeding on leaves with no tannin. When development slows, the herbivore is more likely to be eaten by its natural enemies.

According to one set of theories,[17] plants that are short-lived or well hidden in the community should employ toxins as a defense since they are most likely to be attacked by generalist, nonadapted herbivores. Generalists are species that are fairly nonselective in their food choices, eating plants in order of their abundance in the community. Generalists, because of their relatively wide dietary breadths, are more susceptible to the plant toxins that specialized herbivores have been forced to overcome. On the other hand, such obvious plants as major components of the community or longer-lived species are likely to be found by herbivores. In this case, the herbivores are more likely to be specialists that have evolved resistance to these plants' toxins. Specialists are organisms that over evolutionary time have developed means of sequestering or detoxifying a plant's defensive compounds. Because of this fairly narrow set of adaptations to avoid these toxins, such insects feed on only a narrow range of closely related host species, or even on a single plant species. The ability to overcome a chemical defense, and then feed on a plant that no other species can eat, opens up a new resource for the specialist herbivore. Hence, it would be expected that obvious plants would devote more energy to digestibility reducers than to toxins. Thus, in a plant community, the ratio of these two types of chemical defenses ought to vary among species with likelihood of attack and the degree of specialization of the herbivore species.

In short, herbivores interact with plants in a variety of ways. Some plants may escape herbivory by virtue of being well hidden. Others may be highly toxic, which then exerts selection pressure on herbivores to overcome this defense. Still other plants are visible, easy to find, and therefore likely to be eaten to some extent. Since herbivory is inevitable, these plants may act to minimize the effect by producing digestibility reducers, which, by reducing the availability of nutrients to the herbivore, slow its developmental rate. Extended larval stages in insects can significantly increase their vulnerability to predators. Of course, species such as grasses in prairies that are bound to be grazed have evolved certain traits that allow them to survive grazing by large mammals.

Plant-herbivore interactions can result over time in "coevolutionary arms races."[18] As the insects evolve specific detoxification mechanisms, the host plants evolve modified chemical structures in their defensive compounds. Neither is likely to win, but the race never ends. A similar race takes place with plants and their pathogens. As pathogens evolve to counteract one form of plant resistance, other forms are favored.

Physical defenses in wild plants include spines or thorns (in cacti and honey locust, for example), thick seed coats or fruit coverings (in banana and avocado), and irritating hairs (in the stinging nettle). The advantages of these defenses against herbivory are obvious, although as with chemical defenses, an animal that can counteract such physical defenses has a new food resource available to it.

Recall that natural communities are made up of mosaics of different patches. This spatial heterogeneity allows some plants to avoid insect pests and diseases simply by being hard to find. Plants can essentially hide among the diversity. Wild plants rarely occur in large, homogeneous populations, but tend to be unevenly distributed in space and time. Insect pests and diseases generally increase with host plant population density.[19] An increased concentration of its host plant makes it easier for pests both to locate the host population initially and to move from host to host once established. Thus, smaller patches of a plant species are less prone to disease than are larger patches.[20] Long-range seed dispersal protects a plant's seed-lings by distributing them randomly and unpredictably across the landscape. It is impossible for a pest organism to locate the offspring of a plant with long-range dispersal just by locating the parent plant. Plants typically disperse their seeds over a lengthy period by means of wind or water. The significance of this is that it helps create spatial and temporal heterogeneity so that seedlings may escape herbivores. It may also protect the seeds while they are still on the plant by providing a less concentrated source.

Lastly, availability of host plants to pests may be unpredictable in time. It is better for seeds of a parent plant to germinate across a season or number of seasons in order for some to escape pest attack. Typically, wild plants show staggered dormancy, meaning that a batch of seeds from a single wild plant will not all germinate and develop at the same rate.[21] Heterogeneity in growth rate, phenology, development, and germination enables host populations to persist despite the presence of pests.

These last two forms of escape are quite difficult to exploit in an agroecosystem, where it is important for the farmer to have uniform germination and maturity.

Birds and insects can benefit plants in two major ways: by spreading pollen and seeds and by eating the insects that damage plants. When birds move pollen and seeds around, they spread a plant's genes to mates or to new habitats where offspring may be able to grow. Predators eat insects directly, and the victims are killed quickly. Parasitoids, which are commonly small wasps, lay their eggs within the host individuals. The parasitoid's young develop within the host, ingesting it slowly but not killing it immediately.

High densities of these beneficial predators are favored in mature, stable ecosystems with sufficient species diversity to provide refuges and alternate foods. Beneficial insects and birds need diverse communities where they have alternative food sources all year long. Secondly, vertical diversity—in other words, a mixture of short and tall perennial plants—provides a variety of nesting and feeding sites. An example of a habitat with the right features would be a mature forest with plants that flower throughout the growing season, providing nectar and harboring alternative insect prey, and with a low enough frequency of disturbance to allow predator populations to be maintained. For example, a forest that provides fruits in summer and fall can maintain resident birds that eat herbivorous insects earlier in the year. In contrast, frequently tilled agricultural fields go through long periods of the year with little or no alternative foods for predators. In "Predator Activity and Predation in Corn Agroecosystems," Gerald E. Brust and colleagues found that such agroecosystems simply do not support large populations of such predatory insects as ground-dwelling beetles and spiders.

The diversity found in natural plant communities confers efficiency in resource use and recycling, stability in the face of natural disturbance, resistance to pest and disease outbreaks, and maintains healthy levels of predatory animals that keep herbivory in check. Natural communities are not static collections of climax species, but are dynamic, always changing in response to fluctuations in population size, disturbance regime, and annual climatic changes.

AGAINST THE ECOLOGICAL GRAIN: MODERN AGRICULTURE

In terms of return on labor, North American–style agriculture is undoubtedly a highly productive form of seed, fruit, and fiber production. One estimate holds that one U.S. corn farmer can sustain energetically 380 people per hour worked.[22] As discussed in Chapter 1, however, other estimates show that over the last few decades it has taken increasingly more energy to produce a ton of grain with industrial farming methods. An analysis by Amory and Hunter Lovins and Marty Bender places the ratio of energy expended to food energy consumed in the United States at 9.8:1. But, as will be explored shortly, this productivity has arisen by simplifying ecosystems and tailoring them to maximize a single component, yield, while ignoring or disrupting those factors that act to stabilize and maintain biological communities. Many of the links between organisms, the soil, and the physical environment that serve to regulate natural communities have been decoupled as, instead, agricultural systems have been maintained through massive infusions of fossil fuel–derived inputs.

Other systems around the world are more productive per unit of land area, per unit of investment, or per unit of renewable resource. For example, the ratio of kcals (calories) produced to kcals expended for maize (corn) production with hand labor in Mexico is 12.5:1. For industrialized corn production in the United States, however, the ratio stands at 2.9:1, a process that is 77 percent less efficient energetically.[23]

It has been shown earlier that where natural ecosystem processes break down, energy and material subsidies are needed to maintain system integrity and productivity. In short, modern agricultural methods, while highly productive in the short term as a result of extensive fossil fuel subsidies, are inherently unstable and necessarily short-lived. From earlier discussions of natural ecosystem processes, it can be seen that some of the sources of this instability are biological in origin. This section examines the dynamics within conventional agricultural systems and how they differ from the characteristics of natural ecosystems. Here, the ecological reasons for many of the environmental problems discussed in Chapters 1 and 2, and the need for an ecosystem perspective in moving toward a sustainable agriculture, become evident.

AGROECOSYSTEM PROCESSES

Agroecosystems based on annual crops function as ecosystems arrested in early succession. Thus, they do not benefit from the many processes that stabilize and maintain natural ecosystems. As immature, frequently disturbed ecosystems, they do not display such natural characteristics as efficient nutrient cycling, high species diversity, resistance to pest and disease outbreaks, and resilience to disturbance. Immature ecosystems have low amounts of organic matter, small organisms with broad niches and short life cycles, open mineral cycles, an undeveloped detritus community, poor nutrient conservation, and linear food webs.

Energy. In addition to the solar energy necessary for photosynthesis in an agroecosystem, Chapters 1 and 2 explained that traction, fertility, and pest control take a great deal of fossil fuel energy. The food industry uses additional energy for drying, processing, and transportation. The energy expended in the field is essentially used to fight the natural tendency toward succession and to maintain the agroecosystem artificially at an early successional stage. This is done by creating disturbances (by plowing) that are extensive and frequent enough to maintain the initial conditions for maximum crop growth.

Industrialized farms that grow grain to be fed to livestock for meat production are particularly inefficient when considering how energy is transferred through food chains. Recall the average 10 percent efficiency of conversion of biomass at each step of a food chain. This suggests that exclusively grain-eating human populations could subsist on 10 percent of the arable land required by a food production system that feeds grain to livestock for meat. Here the ecosystem perspective instructs us that much of our soil loss, chemical contamination, and food supply problems could be solved by eating lower on the food chain.

Nutrients. In contrast to natural ecosystems, in conventional agro-ecosystems nutrient cycles are at anything but a steady-state equilibrium. Typically, nutrients flow one way, from their industrial source as chemical fertilizers to harvested materials or groundwater. Fertilizers, nitrogen in particular, are usually applied in large annual pulses, with great potential for inefficient uptake and, consequently, great potential for losses via leaching and erosion.

From what was explored earlier concerning the characteristics of early succession, it is apparent that this earliest seral stage is characterized by inefficient nutrient capture and energy use, overlapping and unfilled niches, and vulnerability to irreversible soil and nutrient loss. The farmer's synthetic fertilizers hinder the activities of nitrogen-fixing organisms and detritivores within the soil because biological nitrogen fixation is favored only where soil nitrogen fertility is fairly low. Open nutrient cycles are essentially one-way flows: nutrients enter from outside the farm and then leave the system, with little internal recycling. This type of flow would be fine if the outside fertility supply were inexhaustible and the downstream effects of nutrient loss minor. Unfortunately, neither is typically true.

Where legumes are not part of a crop rotation, nitrogen enters the system as applied fertilizer, usually as anhydrous ammonia. The ammonium (NH_4^+) is oxidized quickly to nitrate (No_3^-), with the accompanying loss of hydrogen ions (H^+). Over time, the accumulation of these H^+ ions can acidify the soil greatly. Excess nitrate is not held well by negatively charged soil particles and so tends to leach downward and enter and contaminate groundwater supplies.

Agricultural systems generally lose carbon. Plowing coupled with nitrogen fertilizers favors decomposition and oxidation of soil carbon. Erosion of the surface layers of soil also results over time in a net loss of carbon (see discussion under "Soil" below).

Phosphorus and potassium are also commonly applied as inorganic fertilizers. These nutrients are subject to the same losses via erosion as are nitrogen and carbon.

Natural processes that favor nutrient retention and transfer among perennial plants and soil organic matter are not found where plants are annual, the soil is bare for part of the year, diversity is low, soil organic matter has been depleted, and the soil community is depauperate. Our modern agricultural practices break open nutrient cycles so that farmers must intervene and supply lost nutrients to prevent agroecosystems from running down.

Effects of Perturbation on System Stability. News reports of droughts, floods, and disease and insect outbreaks demonstrate that modern agricultural systems are not particularly resilient in the face of climatic or biotic change. The lack of a diversity of species, with each occupying a unique niche, means that the self-damping, nutrient-conserving properties that stabilize natural communities are not present. From the earlier discussion of ecosystem stability, it can be concluded that low stability in agricultural fields is due to more than simply low species diversity, however. Artificial selection of cultivars has eliminated many defensive traits present in wild plants. Also, frequent cultivation disrupts or eliminates many members of the soil community, which, as has been shown, are the foundation for healthy ecosystem function.

Soil. Many of the ecosystem processes that tend to hold soil in place—for example, a diverse community of perennial vegetation and a healthy soil community—are broken down when soil is repeatedly tilled. During some part of each year, roots to hold soil and a plant canopy to protect the soil surface from water and wind erosion are absent.

Several marked changes occur when virgin soil is cultivated. Organic matter content rapidly decreases. The large pores, so crucial for soil function, are destroyed; changes in soil physical properties increase the risk of erosion; and invertebrates, especially earthworms (*Lumbricus* sp.), are suppressed.[24] Continuous cultivation leads to greater rates of leaching. Breaking ground unbalances nutrient cycles by stimulating the activity of soil organisms, which leads to net loss of organic matter and release of soil carbon into the atmosphere as carbon dioxide. Soil devoid of organic matter holds moisture, oxygen, and nutrients poorly.

When virgin soil is cultivated, organic carbon, nitrogen, and phosphorus decrease from the surface horizons. One study found that cultivated soil of shortgrass prairie contained only 58 percent as much mineralizable nitrogen as undisturbed soil and less than half the microbial biomass in the top 15 centimeters (equivalent to 6 inches).[25] In another study, cultivation reduced the average amount of carbon tied up in microbial biomass by 62 percent and nitrogen by 32 percent.[26] In yet a third study, nitrogen losses from the A horizon over about forty years of cultivation in western Kansas averaged about 40 percent.[27] As early as 1957, W. V. Bartholomew and associates, in an article in the *Agronomy Journal*, documented that continuous corn cultivation in Iowa had deprived the soil of its nitrogen reserves. Cultivation also leads to a marked loss of organic carbon and a change in fine texture. H. J. Haas and associates, in a USDA publication, reported that organic carbon was 26 percent less in soils cultivated for at least fifty years. Similarly, in an article in *Biogeochemistry*, Schoenau and colleagues reported their findings that seventy years of cereal-fallow cultivation of prairie soils reduced the concentrations of total phosphorus in the A horizon by 19 percent to 25 percent. Most of this loss occurred in the organic fraction. Cultivation also allows excess water to move through the soil, thereby leaching soluble nutrients from the profile. Clearly, cultivation has profound consequences on the living soil community and its ability to hold and recycle nutrients.

As discussed in Chapter 2, erosion is a serious problem, perhaps the main conservation problem on about half of all cultivated U.S. cropland.[28] Severe soil erosion leads to irreversible soil damage. The soil profile holds the key to a soil's vulnerability to erosion. In cultivated soils, the A horizon is important in controlling water, heat, and gas balances. Plant roots and available nutrients are concentrated in this layer, whose depth may vary from a few centimeters to 50 centimeters (up to 20 inches). Under cultivation, the A horizon typically decreases in depth or is removed by erosion, or, when tilled, mixes with the B horizon. Organic matter is one of the first soil constituents removed by erosion. Its high concentration in surface soil and low density leave it particularly vulnerable to both wind and water erosion. Its loss directly affects nutrient cycling in two ways. One is an immediate loss of nutrients associated with the organic matter. Second is the loss of much of the soil's cation exchange capacity, or capacity to retain nutrients. Not only is the current supply of nutrients removed, but the remaining soil is less able to store incoming nutrients.

The permeability of soil to water is controlled typically by the B horizon, or subsoil. As the clay-laden B horizon gets closer to the surface, soils drain poorly and run-off, and hence erosion, increases. Exposed subsoil may be unfavorable for plant growth in terms of pH and high concentrations of iron and aluminum. Thus, plant productivity is modified by soil profile characteristics and the distribution of nutrients with depth. Erosion removes disproportionate amounts of carbon, nitrogen, and phosphorus because these nutrients are concentrated near the surface. Furthermore, because some soil attributes are irreplaceable, erosion rates alone may not be good indicators of soil degradation. Conceivably, soil may be destroyed faster than it is eroding.

Even short intervals of exposed soil can result in major erosion if the timing corresponds to times when windstorms and rainstorms occur. For example, it has been estimated that 50 percent of erosion on corn land occurs during heavy rains in May and June, when the crop is just emerging and the soil is not protected by a crop canopy.

Time Factor, Succession. Annual crops are analogous to early colonizing plants in that they're bred to exploit ephemeral resources and maximize seed output. Agricultural fields essentially mimic the first stage of an ecosystem following a general devastating disturbance. Repeated ground preparation favors annual crop plants that grow rapidly, but it does not provide a means of retaining soil or efficient nutrient, light, and water use. A more diverse community of plant species with complementary niches is required to make the ecosystem function more efficiently.

POPULATION AND COMMUNITY PROCESSES

In agroecosystems diversity is low—indeed a monoculture often—which means all individuals in the population are using the same resources simultaneously, and there is no room for divergence of resource use by occupation of different niches. Therefore, mineral resources available when crops are not actively growing are subject to leaching or other forms of loss. Low plant-species diversity in the field has ramifications along food chains as fewer beneficial organisms can be supported. Moreover, many predatory insects and birds are destroyed through the wholesale application of biocides. Hence, the farmer with synthetic chemicals assumes the roles played by predatory insects and vertebrates in controlling pests and by competitively dominant plants that eventually shade out or otherwise exclude weedy species in natural communities.

Size Distribution and Genetic Diversity. Many of the pest problems observed in monoculture agriculture relate to its extreme uniformity in niche, chemistry, and genotype. In the ideal industrialized crop field, all individuals are the same size and age and use resources simultaneously. They present a chemically and physiologically uniform and predictable food source for pathogens and herbivorous insects. They are even phenologically uniform, meaning that the whole population is in the same life cycle stage simultaneously. Therefore the crop is unable to benefit from the protection that genetic, size, and age diversity affords plant species in natural populations.

Artificially selected, high-density monocultures of nearly genetically uniform crops provide an ideal resource for pathogens. Because pathogens commonly are highly host-specific, they are capable of rapid epidemic increase in single-species fields. In agriculture, human control of the composition of host populations predisposes such systems to sporadic, often devastating epidemics. Even if the crop has been bred for resistance, this trait may be controlled by a narrow set of genes identical in each individual. As stated earlier, this imposes a strong selection pressure on a pathogen to overcome this defense. The fewer the genes for resistance, the faster the insect or disease can evolve the ability to overcome the resistance. Since the attacker generally has a shorter generation time, and thus potentially faster rates of evolution than even annual crop plants, it can overcome this resistance. This is what keeps crop breeders employed. They must continually develop new varieties with new forms of resistance. On the other hand, genetic variability for resistance, which is more typical of wild plants, is much more difficult for the pest to overcome.

Mature Communities as Mosaics. From the perspective of herbivores, a monoculture of crop plants is a barrier-free feast. From the weeds' perspective, it is an unimaginatively predictable opponent. The monoculture benefits from none of the advantages of heterogeneity inherent in naturally patchy communities. Insects dining on a crop plant can easily find a second course next door, whereas in a natural community they would have to spend some of the energy that they just consumed searching for the next plant to eat. In most natural ecosystems, weeds are unable to compete well against established perennial vegetation, and they therefore play only a minor role in nature. They maintain a toehold in the community via local disturbances, after which they may quickly colonize and hold exposed soil and nutrients. In contrast, weeds compete with crops to colonize bare ground each year, and consequently the role of weeds in annual cropping systems is much greater. Their impact may be the biggest impediment to agriculture in some regions.

Niches. Unfortunately, most annual crop plants occupy the same niches as weeds. Because they use the same part of the soil at the same time, they tend to compete directly with weeds. In a monoculture, the crop can fill only one niche. In such a depauperate community, weeds have many opportunities to invade. It is not even clear that all weeds are actually harmful to crops. Weeds filling complementary niches to crops could be beneficial, as noted above. Clearly, in a monoculture there is little if any potential for the positive, beneficial interactions that occur among different species in natural communities.

Competition from Weeds. In a monoculture, competition occurs in its most intense form. Here competition is mainly among crop plants, all of the same species and genetically very similar. Since weeds occupy similar niches to crops, competition between crop plants and weeds is also intense. In a monoculture, all the plants scramble to procure the same resources at the same time.

Facilitation. In modern agricultural fields, facilitation is generally minor, except where legumes are used as part of a rotation. Nitrogen fixation is usually inhibited by previous use of synthetic nitrogen fertilizer, making it difficult for legumes to survive in agricultural fields in the face of more vigorous species. Frequent or chronic disruption of the soil community inhibits the establishment of the types of mutualistic links, such as those between plants and mycorrhizal fungi, so prevalent in natural ecosystems.

Insects and Diseases. Although farmers generally plant single varieties of crops so that whole fields have uniform resistance, this extreme homogeneity is usually not a natural state, even for the ancestors and relatives of agricultural crops. Surveys of wild relatives of crops demonstrate that natural populations may harbor a substantial variety of forms of resistance to pathogens.[29] The potential is present in crops' genetic background to mimic this natural and effective form of disease resistance.

In the process of domestication, several defense mechanisms against pests that are common in wild plants were deliberately eliminated. Chemical protection, physical protection, seed dispersal, and asynchrony of maturation and germination have been bred out. Although these factors defended plants from natural pests in the wild, they also made them unsuitable as crops.

Chemicals that protect wild plants are particularly undesirable in crops. Many of these defensive compounds taste bad or are directly toxic to humans and livestock, which explains why they were bred out. For example, many wild members of the squash family are spiny and taste bitter, and seeds of many wild legumes contain digestibility-reducing compounds. Thus, many chemical defenses are basically unavailable to crop plants.

Seed dispersal is a major way for wild plants simultaneously to protect their seed from animals, colonize new sites, and minimize competition and disease spread among offspring. The propensity of wild fruits to open, or shatter, and present their seed for dispersal was bred out of crop plants in order to increase yield. Again, the farmer desires uniform stands for ease of cultivation and harvest and so has selected for uniformity in germination, elimination of seed dormancy, and uniformity in growth and maturation. As with defensive chemicals, the capacity for some types of biological crop defense is in direct conflict with agronomic suitability. Whenever a natural defense is bred out of a domesticated crop, the farmer must step in and provide that protection either chemically or through management practices. An open question within sustainable agriculture research is to what extent it is possible to allow and rely upon natural defenses in crops.

A move toward a sustainable agriculture will involve shifting the dominant goals from industrial productivity and efficiency to goals that acknowledge agriculture as a biological and social process. Some examples of the application of ecological philosophies to agriculture are explored in Chapters 4 and 5. Although an ecological perspective has much to offer sustainable agriculture, agroecologists must consider not only the biological and economic factors in their formulations, but also the social realm within which a sustainable agriculture is to develop. This topic is considered in Chapter 6.

This chapter ends with a summary of some important ways in which conventional agricultural ecosystems differ from natural ecosystems. Table 3.1 lists some of the sustainable features of natural communities that are absent from conventional agriculture. One feature necessary to create a more sustainable system of agriculture is a higher degree of resilience, or ability to weather climatic or market extremes. Natural ecosystems have shown that, to a large extent, resilience can come about through high species and genetic diversity, tight nutrient cycling, and the interdependence of species whose niches complement one another in space and time. Ultimately, as agriculture moves toward sustainability, a higher percetage of its energy will come from the sun and more of its nutrient inputs will be derived from local, replaceable sources.

TABLE 3.1

Comparison Between Conventional, Industrial Agriculture and the Native Prairie Ecosystem for Some Factors That Contribute to Sustainability

	Conventional agriculture	Native prairie
Fragility	high	low
Resilience	low	high
Species and genetic diversity	low	high
Rate of nutrient flux	high	low
Degree of biotic interdependence	low	high
Energy source	solar and fossil fuel	solar
Nutrients	from fertilizers	local, recycled

4

A New Agricultural Perspective: The Case for Ecological Agriculture

THE IMAGE OF a modern corn field is that of a neat, geometric box with sharply creased edges, hard corners, a crew-cut-even top, and straight rows cutting across bare, exposed soil in neat green and brown stripes. It is a crisp, brittle image, pleasing perhaps to eyes with an industrial bias.

The image of a natural ecosystem—the prairie, for example—is that of a fuzzy blanket, with an intricate design of many hues, tightly woven but fraying at the edges so that it is without sharp boundaries, and interwoven with the fabric of the earth that it covers protectively. It is a soft, resilient image, pleasing perhaps to eyes with an ecological bias.

The two images bear little likeness, except for overall greenness. The second image is chosen here as a model for imparting nature's steadiness, resilience, and sustainability to agriculture. How can the transition be made? How can agriculture be transformed? This chapter describes a possible path for that transformation.

Presented first are several examples—integrated pest management, intercropping, and conservation tillage—where agriculture has successfully incorporated selected ecological processes extracted from nature. Next, the chapter's central theme is introduced and developed: that local ecosystems provide the most appropriate structural models for agriculture. By mimicking a natural vegetation structure, farmers can copy a whole package of patterns and processes that have developed and worked in an ecological or evolutionary time frame. With this structural approach, a multitude of beneficial ecological processes can be incorporated into agroecosystems. This orientation departs radically from the narrow single problem/single solution approach typical of agricultural science. Examples of agricultural systems based more or less on this concept are found around the world, and several are reviewed here: Native American agriculture in warm deserts, temperate agroforestry, tropical successional agriculture, and tropical agroforestry. Finally, a proposal is presented for the North American grain belt: polycultures of herbaceous perennial seed crops designed to mimic native prairie ecosystems.

NATURAL ECOSYSTEMS AS MODELS FOR SUSTAINABLE AGRICULTURE

Farmers in the present industrial agriculture mold play the role of ecosystem simplifiers. By applying various biocides they attempt to eliminate such complicating factors as weeds, insect herbivores, and fungal pathogens from their fields. By applying synthetic nutrients they attempt to short-circuit biologically driven nutrient cycles and control soil fertility. What remains of the biotic community they direct to a solo peak performance to maximize short-term yield of products for human use. This single-species star performance comes at the expense of other ecosystem components. As shown in Chapter 3, biological integration among species is low in such simplified communities, and uniform crop populations are especially vulnerable to insect and disease outbreaks. Thus farmers create a fragile system when they simplify the agroecosystem so drastically.

To achieve the resilience of natural ecosystems, a new model for agriculture is needed—one that uses time-tested local ecosystems as models. This requires a change in perspective from the level of a population of crop plants to the ecosystem level of biological consideration. It requires a broadened means of accounting that includes assessment of the status of the soil, noncrop organisms, and water in the measures of success of agricultural practices. It requires more emphasis on the interconnections among species and on the interdependence of species and their physical environment. It requires that greater weight be given to factors other than maximum production of a single crop. It means that a farmer's role will be more akin to that of an orchestra director than to the ecosystem simplifier's role as a solo-ist's coach.

Natural communities have been tailored by climatic and evolution-ary forces to accommodate particular environments and to endure. They provide the best examples of the characteristics necessary for sustaining an agriculture that neither depletes the environment nor depends upon exhaustible resources. Most natural ecosystems char-acteristically have low levels of soil loss, diverse assemblages of spe-cies coexisting in dynamic equilibrium, dominant plant species that are perennial and adapted to local soils and climatic patterns, an ex-clusively solar source for energy, internal recycling of nutrients, net production equal to respiration so that total biomass is at steady state (neither increasing nor decreasing over time), high efficiency of en-ergy transfer along food webs, and resistance to native herbivores and disease. Just how these characteristics are worked out varies a great deal from one region to another, and thus the shape, structure, and appearance of ecosystems vary drastically too, from prairie to forest, from desert to marsh. Still, the native community generally manifests these same sustainable features. These natural plant com-munities constitute the best structural fit to their native region and have much to teach about how to farm sustainably.

In contrast to the prevailing approach that imposes industrial ag-ricultural visions upon resistant landscapes, the natural community presents the most suitable model for agricultural practices appro-priate to particular landscapes and climates. Using nature as a stan-dard, model, or analogy for agriculture is a notion difficult to articu-late, however. What is meant by "nature" is a debate perhaps best left to philosophers. For the purposes of this book, the entire package of a natural ecosystem's stable vegetation structure and the stabiliz-ing processes woven into it are considered to make up "nature's model." To apply nature's model in the process of developing an ag-riculture for a particular region, it should first be asked whether ag-riculture is an appropriate activity for that region. The resilience of the native ecosystems of that region provides the key to answer this question. Certain ecosystems, such as those too fragile or too harsh to farm sustainably, should never be converted to agricultural uses. If it is decided that a landscape can tolerate agriculture, it is then nec-essary to look to the native ecosystems to see what types of agricul-ture will work. It is necessary to understand what makes those eco-systems work sustainably in that particular landscape. Researchers might ask: Why is the landscape here covered with trees? Why is this ecosystem dominated by drought-tolerant succulents? Why do C_4 grasses predominate here and why are trees scarce?

BLENDING THE NATURAL INTO AGRICULTURAL ECOSYSTEMS

Of course, agricultural and natural ecosystems are related biologically, and some agroecosystems are more natural than others along the nature-agriculture continuum. Despite the best attempts to industrialize and sterilize it, field agriculture remains an inherently biological process. As with wild plants, the growth of crop plants responds to soil, climate, and other organisms. Yet room remains for more of the natural to be incorporated into conventional temperate-zone agriculture.

Although it is an oversimplification, it may be a useful exercise to contrast natural and agricultural ecosystems. Natural ecosystems provide low seed and fruit production for human use, but are sustainable barring human disruption or major disasters. Conventional agroecosystems produce high yields of particular products, but are necessarily ephemeral where they diminish soil and water and depend on fossil fuels. When human populations were smaller, societies could subsist on what wild ecosystems provided. This often meant low population density, a nomadic lifestyle, and a diet that shifted seasonally.[1] Maple-sugaring in New England and rubber-tapping in South America remain as two examples of modern gathering approaches. The harvester gathers a useful product from the environment, but without depleting the ecological capital or altering the species composition of the natural community. Present human population densities, concentrated in cities and removed from their main sources of food production, preclude a solely hunting-and-gathering lifestyle. Some indigenous cultures, however, have not experienced the nature-agriculture split so clearly. Indeed, the hope for sustained food and fiber production lies in a scientific blending of the natural patterns and processes that stabilize natural communities with agricultural needs.

So how can such a blending be achieved? How is it possible for agricultural systems to incorporate such processes as tight, closed nutrient cycles and biological control of pests? People have explored different avenues to achieve this goal. For example, practitioners of organic agriculture apply intensive and careful management to achieve biological control of insects and soil fertility. These practices are largely process-oriented, referring to agricultural practices that attempt to employ some ecological process that occurs in nature without imitating the vegetative structure of a natural community. For instance, nitrogen supply may come from legumes in rotation. Crops are still grown singly, in monoculture, not in mixtures, as would be found in natural communities. The process of legume nitrogen fixation is imported apart from the mixed species culture.

What is proposed here is a different tack, one that uses a mostly unexplored structural model to achieve sustainable processes within agricultural fields. In short, this structural model imitates the vegetative structures of natural communities in agricultural designs. By using nature as a physical model and designing agriculture "in nature's image," this structural approach attempts to incorporate a whole array of ecological stabilizing processes into the human-managed mimics. In this model, biological nitrogen supply would come by including legumes in a mixture of crop species in similar proportions to their counterparts in native communities.

The structure and process models do not constitute a perfect dichotomy. The structural model embodies the process model, but the reverse is not necessarily true. The process model usually involves transplanting one or a few ecological processes into an otherwise fairly conventional agriculture. The structural approach is more holistic in that it imitates an entire working community or ecosystem rather than dismembering a system to try to understand all the individual processes. When the whole structure is imitated, certain emergent properties that are expressed only at the ecosystem level can appear in agroecosystems.

EXAMPLES OF ECOLOGICAL PROCESSES INCORPORATED INTO AGRICULTURE

Those working in the direction of a permanent agriculture by blending ecology and agriculture have already incorporated some important processes characteristic of natural communities into agroecosystems in various ways. Three of these practices—biological management of detrimental organisms, intercropping, and conservation tillage—incorporate aspects of nature as a functional model for processes leading to sustainability and have clearly shown some environmental benefits. Following is a description of the ecological benefits and shortcomings of these process-oriented models and an explanation of how each is represented or might be represented in agriculture.

Integrated Pest Management. Biological control, and various forms of integrated pest management (IPM), are based on an understanding of predator-prey and host-parasite interactions and insect life-history patterns in nature.[2] The IPM movement arose in the early 1970s in response to concerns about impacts of pesticides on the environment. By providing an alternative to the strategy of unilateral intervention with chemicals, it moves from a brute force, industrial mode of pest control to a stance that is more ecological and humane. To some extent, then, IPM changed the philosophy of crop protection to one that entailed a deeper understanding of insect and crop ecology and relied on the use of several complementary tactics.

Integrated pest management incorporates several diverse tactics of pest control, relying first on natural control factors (pathogens, parasites, predators, and weather, for example) and management. It relies on pest population dynamics, such as length of immature stage or reproductive period, to suggest exploitable weaknesses in the biology of the pest. Another large part of IPM involves determining the economic thresholds of yield and refraining from insect control below these thresholds. The IPM practitioner evaluates whether there will be sufficient pests to justify control by the grower, whether the pests will last long enough or remain dense enough to lower yields, and whether natural controls will intervene. The actions taken may be cultural methods, biological controls, the use of toxic chemicals, or a combination of these. Cultural methods include manipulation of the density and diversity of vegetation, cultivation, sanitation, variation in planting and harvesting dates and the varieties planted, and alteration of fertility or irrigation levels. Classical biological controls may employ predators, parasites, pathogens, and nematodes and may involve foreign exploration to find natural enemies, the mass release of enemies, and the conservation of natural enemies. Chemical methods employ substances that disrupt the functioning of behavioral pheromones (chemicals emitted by insects that facilitate orientation and mating) and judicious use of pesticides as a last recourse.

Integrated pest management may sound complex compared to the simple application of chemical biocides, but it is actually quite simplified compared to the integrative pest-management processes within natural ecosystems. In nature, the dynamics of herbivore populations are influenced by the structural diversity of surrounding vegetation as well as by the population states of predators and competitors. Typically, however, IPM strategies are applied in simplified systems so that this complex ecosystem structure is not imitated. One or two aspects of a pest's biology are extracted from nature and imported to the agroecosystem. Such aspects of an organism's biology might include its population dynamics or the relationship between that particular pest and a particular predator. In nature, interactions between species exist as components of the larger community and are much more complex. A simplified approach considers neither the complexity of the ecosystem in which the natural interaction takes place nor the complexity of the agroecosystem for which the control is intended. Thus, many introduced biological control agents fail to perform as expected due to unforeseen elements in their new environments.[3]

Because of this limitation, IPM has unfortunately not changed enough of the fundamental industrial structure of modern agriculture. As a concept, IPM has been incorporated into large-scale farming to reduce the costs of pest control. Although pesticide applications have become fewer and more efficient, the use of toxic substances is still maintained. Hence, because IPM remains too chemical-dependent, and too wedded to such simplified, narrow solutions as a single predator or herbivore species to combat a problem pest or weed, it is only a partial solution to the problem of agricultural sustainability.

Intercropping. The practice of raising two or more crops in the same field at the same time is called intercropping. It turns out that this relatively simple step away from monoculture has many benefits. At first glance, intercropping may appear to be more of a structural than process-oriented model, but in reality most intercrops have been designed to reproduce a particular ecosystem process, rather than to imitate ecosystem structure. As a result, most intercrops bear little resemblance to the natural ecosystem they replaced. For this reason we include intercropping with process-oriented models.

Intercropping enhances the efficiency of land use. It takes advantage of the complementary aspects of species' niches. Species that overlap minimally in timing, spacing, or type of resource use have most often demonstrated a potential to overyield when grown together—that is, they yield more together than when grown separately. This occurs because differences among species in patterns of resource use enable mixed-species fields to use available resources more thoroughly than any one species can. In fact, multispecies fields typically yield more per unit area than do equivalent areas planted to monoculture.

Of course, not all mixtures of crops overyield. One difficulty in designing intercropping systems is that it is not always possible to predict how a crop will behave in polyculture from its performance in monoculture. For example, some plants change their patterns of nutrient uptake when grown in association with different species.[4] Or a prostrate plant, vigorous in monoculture, may be shaded out by a taller neighbor. When designing optimal crop mixtures it helps to have an understanding of the patterns of community structure and niche partitioning in natural communities. For example, a good intercrop plan might mix plants with different root morphologies or seasonal growth patterns. Legumes are especially important as intercrops for providing nitrogen to the cropping system. (An exhaustive review of intercropping systems is not provided here since several excellent reviews are available.[5])

In addition to the obvious overyielding advantage, mixed-species cropping offers benefits in the management of detrimental insects, diseases, and weeds. Intercrops are usually designed to tap and incorporate particular processes, such as predator attraction, inhibition of insects' host-finding behavior, and trap-cropping (growing plants that are favored by pests along with crop plants so that pests are attracted away from crops) to minimize insect pests.[6] The kind of species and structural diversity that confers stability in natural plant communities can, in agricultural fields, increase densities of natural predators and reduce rates of pest colonization.[7]

Similarly, mixtures of crop species, and of genotypes within crops, create patterns that increase the genetic diversity of crop fields and thereby can reduce the incidence and spread of some plant diseases.[8] Oftentimes, a mixture of genotypes expressing different forms of resistance to a pest or disease is more effective in resisting depredation for a longer period than is a monoculture of individuals, each with identical genetic resistance. Because the life cycles of most microbes and insects are considerably shorter than those of their host plants, these pests are generally able to evolve the ability to overcome a plant's defenses faster than the plant population can evolve a new form of resistance. Incorporating a variety of resistance types into a crop population, then, can limit the pest's ability to overwhelm the entire crop. Similar stability can be conferred to agroecosystems in which members of a genetically diverse crop population express resistance to a variety of different diseases likely to be encountered.

In addition, the overyielding effect may be more pronounced in the presence of disease for two reasons. First, susceptible members of mixed stands are less damaged than are those in pure stands. Second, resistant individuals will oftentimes compensate with yield increases after their neighbors fail.

Intercropping can also help control weeds. For instance, one or more crop components may be an effective allelopathic species. Allelopathic plants release substances from their leaves or roots that directly, or indirectly by manipulating soil organisms, inhibit the germination or growth of immediate neighbors. Many wild and cultivated plants are reputed to be allelopathic.[9] Another important way in which weed growth can be reduced in polyculture is through continuous shading of the soil surface by perennial plants or via relay cropping.[10] Weeds are less able to compete where most of the soil surface is shaded by crops for most of the growing season.

Intercropping is a step in the right direction toward diversifying, and thus stabilizing, agriculture. Most intercrop systems, however, are composed of annual crops. These conventional crops still require high energy inputs and are often incapable of adequately protecting the soil from unsustainable levels of erosion. Although farming systems that incorporate biological, on-farm sources of fertility—through the use of legume intercrops for example—are moving to close the open nutrient cycles that characterize conventional farming, many otherwise organic farms in developed countries still depend too heavily upon fossil fuels or great amounts of subsidized irrigation water. Moreover, because an annual polyculture may bear no structural resemblance to the vegetation type that formerly occupied the landscape prior to its conversion to agriculture, it may exist out of context with that landscape's climatic regime. Annual polycultures may thus be vulnerable to devastation by climatic extremes.

Conservation Tillage. Conservation tillage is an umbrella term that encompasses a variety of practices ranging from no tillage to some reduction in the frequency or depth of plowing. Reduced tillage allows the eventual formation of soil strata that are similar to those of native soils and can enhance populations of such soil organisms as earthworms, insects, fungi, and microbes. Some of the organisms favored in no-till fields include predatory insects that can significantly reduce numbers of pest larvae where the soil is disturbed only minimally.[11] A healthy, active soil community shows relatively high efficiencies of nutrient cycling. Nutrients from decomposing organic matter become available slowly and are thus less vulnerable to leaching or erosion losses. Because soil carbon is conserved rather than oxidized, soil organic matter tends to increase over time.[12] In sum, maintenance of the soil community in a largely natural state can translate into improved crop vigor, lower levels of some diseases and pests, favorable soil structure, and retention of nutrients.

In no-tillage agroecosystems, crop residues typically remain on the soil surface rather than being turned under. Surface litter allows mycorrhizal associations to develop, as the fungi use carbon in the residue for food. As these mycorrhizal associations flourish, soil nutrients, especially phosphorus, accumulate in the A horizon near the surface. The fertility of the topsoil rises and can approach that of native prairie soil.[13] In "Detritus Food Webs in Conventional and No-tillage Agroecosystems," Paul Hendrix and associates have demonstrated an increase in available soil nutrients within no-till fields. Moreover, crop-derived mulch provides a buffer that protects the soil surface from wind and water erosion and increases soil water retention.

Approximately 20 percent of U.S. on-farm energy usage is associated with traction.[14] Thus, any practice that reduces or eliminates tillage should translate into savings in energy and costs for farmers. Unfortunately, because most conservation tillage depends upon high levels of herbicide use for initial weed removal, savings in energy are to some extent offset by additional herbicide costs. Moreover, these herbicides can severely reduce or eliminate populations of beneficial soil organisms, and run-off and leachates of agricultural chemicals can pollute surface and groundwater supplies.[15]

PROBLEMS WITH PROCESS-ORIENTED APPROACHES

All of the practices outlined above are valuable and serve perhaps as necessary and logical intermediate steps in the marriage of ecology and agriculture. As pointed out, however, none of these processes alone is sufficient; ultimate solutions to the problems of agriculture lie beyond what are effectively mere modifications of standard agricultural practices. Cropping systems that involve annual plants may still result in unsustainable levels of soil erosion and net loss of soil organic matter by decomposition. Much conservation tillage relies on herbicides for weed control, and many of these have been implicated in the destruction of beneficial soil organisms as well as the contamination of ground and surface waters. As practiced, IPM guarantees a continuing role for chemical pesticides in modern agriculture. Few organic farming studies, such as those recommending frequent cultivation to replace herbicide use, are addressing the issue of agriculture's fossil fuel dependence or even soil erosion. In short, most current alternative agronomic systems still leave much unaddressed from a long-term ecological point of view. The ecological and economic problems of current farming systems can be resolved only with innovative research approaches.

The problem of sustainability is too often regarded solely as a technological problem of production. This approach limits the researcher's ability to understand and address the fundamental ecological reasons why agricultural systems are unsustainable. The "nature as model" approach provides an ecological rather than a technological orientation to solving the problem of agricultural sustainability. This approach seeks to diagnose the health of an agroecosystem and to discern the systems-level ecological principles necessary to develop a sustainable agriculture. Simply focusing on technological aspects of the problems and seeking quick-fix technological solutions obscure the fundamental problems that lie behind the technology-induced environmental crisis. Even though some promoted alternative technologies may be low-input, without a whole-ecosystem approach, all the interconnections among various problems cannot be addressed.

NATURE AS A STRUCTURAL MODEL FOR AGRICULTURE

The structural model for agriculture advocated here suggests that agroecosystems should mimic the vegetation structure of natural plant communities. This means developing agroecosystems that incorporate crops as structural mimics, and thereby functional analogs, of wild species. Cropping systems then would resemble, and behave like, natural communities. If structural mimics are successful, then such agroecosystems should reflect many of the processes that stabilize natural ecosystems. These include vegetation adapted to seasonal precipitation patterns, tight nutrient cycles, compatibility in resource use among species, soil preservation, and biological methods of crop protection. In this holistic approach, by mimicking nature, one incorporates nature's functional attributes into agriculture, but without necessarily knowing in great detail how each minute particular of the system works. A beauty of this structural approach is that, theoretically, it should work for many different types of ecosystems. Following are some examples that each describe how a particular ecosystem functions as a model, list its sustainable features, and explain how an agriculture appropriate to that ecosystem could or does imitate it structurally. In a number of cases, indigenous agriculture holds the key to some basic agroecological principles in need of consideration in making agriculture more sustainable.[16]

WARM DESERTS

The first example of an ecosystemlike agriculture is based on the North American warm deserts—the Mojave, Sonora, and Chihuahua deserts—located in the southwestern United States and northern Mexico. To discover how these ecosystems can work as a model for a sustainable agriculture, it is first necessary to explore the physical conditions in the deserts and then how the vegetation has adapted to survive these conditions.

The climate of these regions ranges from warm temperate to subtropical. Within these ecosystems rainfall is very low, on the order of a few centimeters per year, due to a rain shadow created by the Sierra Nevada, which intercepts moist winter air masses brought in by the prevailing westerlies. Rates of evapotranspiration are high. Moreover, precipitation is highly variable among seasons and from year to year. In contrast to cool deserts, which have summer storms of high intensity but short duration, most precipitation in warm deserts falls as rain in winter and spring. In the Mojave Desert, winter rains are of long duration, but low intensity. In contrast, the Chihuahua Desert experiences summer cyclonic thunderstorms of short duration but high intensity, which leads to much run-off and little percolation into the dry soil. The climate in these deserts is harsh, yet they support stable vegetation. How do these desert ecosystems function under such harsh conditions?

Warm-desert soils are low in organic matter and have a poorly developed crumb structure.[17] Desert soils are slightly acidic to alkaline at the surface and generally do not have well-developed profiles, but may develop calcium carbonate ($CaCO_3$) accumulations in the upper 2 meters (6.5 feet). A horizon may form a hardpan, a layer called a caliche that is impervious to water, which may be very thick (up to 90 meters, or nearly 300 feet) due to $CaCO_3$, silica, or iron deposits. Whereas in most soils the main carbon fraction is in the form of organic matter from the decomposition of plants and animals, in deserts the ratio of carbonate to organic matter in the soil may exceed 10:1. In addition, many desert landscapes are dominated by dunes, whose soils are sandy and thus allow for rapid percolation and subsequent storage of soil water. Dunes may therefore be among the moister sites in deserts.

Desert soils are very heterogeneous, and this heterogeneity determines the spatial patterning of desert vegetation. For example, the presence of a caliche layer in an area influences which plant species will be present there. Sand dunes also support their own characteristic floras. Areas called playas form where run-off from mountains accumulates in depressions and then evaporates, leaving deposits high in calcium and sodium salts. Alluvial fans are conelike depressions that originate from canyons in mountains and fan out from the mouths into valleys. Moisture and nutrients are concentrated within these alluvial fans, which collect run-off and debris downslope from uplands. Where several fans from adjacent canyons coalesce, an area referred to as a bajada forms. The lower ends of bajadas may have very fine soils. Upper bajadas tend to have larger average soil particles, lower salinity, and support higher plant species diversity than lower portions of bajadas. Certain cyanobacteria, algae, or lichens generate hardened soil crusts on clay or silty soils. These organisms both stabilize the soil surface and may fix atmospheric nitrogen in quantities significant for desert ecosystem nitrogen cycling.

Desert plants display a great variety of strategies for coping with the extreme conditions imposed by desert climates. Annual plants effectively avoid drought by remaining dormant as seeds during periods of extreme drought, then germinating and growing in response to precipitation. Also, annuals may be restricted to the more mesic alluvial fans. Succulents such as cacti are able to store water and photosynthesize with very high water-use efficiency by virtue of their CAM photosynthetic pathway. Some plants show very elegant adaptations to xeric conditions. For example, big sagebush (*Artemesia tridentata*), a shrub, displays a seasonal change in the shape of its leaves. In the spring it produces large, ephemeral leaves that remain until the point of soil moisture stress in the summer and then fall off. Smaller, persistent leaves develop in late spring. These smaller leaves remain on the plant, continuing to photosynthesize throughout the winter. Some plant species form fibrous root systems that efficiently acquire water and nutrients near the soil surface following a rain. Other desert plants, trees especially, develop deep taproots that can feed deep in the soil profile or even tap the water table. Moreover, such leguminous trees as mesquite (*Prosopis glandulosa*) are critical to the nitrogen balance of the desert ecosystem. Many perennial grasses grow during the narrow window of favorable temperature and soil moisture in late spring and early summer. Many perennial forbs can grow for longer periods by storing carbohydrates and then moving these reserves upward from their fleshy roots.

This whole array of desert survival strategies is represented in the vegetation of warm-desert ecosystems. The vegetation of the Mojave Desert consists of low-growing, widely spaced perennial shrubs—for example, creosote bush (*Larrea tridentata*) and white bursage (*Ambrosia dumosa*)—a few cacti, locally common yuccas—especially Joshua tree (*Yucca brevifolia*)—and annuals. The Sonora Desert is characterized by shrubs, subtrees, cacti, and annuals. The Chihuahua Desert, a higher elevation ecosystem, is cooler and wetter than the other two warm deserts. Thus, the vegetation of the Chihuahua Desert consists predominantly of shrubs, *Acacia* trees, *Yucca* species, other small trees, perennial grasses, and annuals.

Sustainable Features. Because of the severity of the climatic conditions, overall productivity in North American deserts is generally low. It ranges from about 30 to 300 gm/m^2 per year (about 270 to 2700 lb/acre per year). Yet, despite low productivity, sustainability of warm-desert ecosystems is conferred by a vegetation that comprises species with marvelous adaptations for acquiring and storing water, avoiding drought, and surviving at low population densities. Desert communities are characterized by great spatial and temporal diversity among species. Different species, even different vegetation types, segregate out across the heterogeneous landscapes. Temporal diversity comes from the many species that are apparent only when water is available, surviving as seeds in the soil during dry periods. Adapted to use water in its season, these species are ephemeral—arising, flowering, and setting seed in response to rain. Thus, the vegetation endures harsh conditions by means of a variety of strategies that enable it quickly to take advantage of scarce, unpredictable resources and to endure long dry spells.

Yet the sustainability of desert vegetation should not be confused with resilience. Desert ecosystems are fragile, and due to their fragility they are particularly susceptible to damage by human activities. Improper irrigation of crops combines with high evaporation rates in deserts to create levels of soil salinity that are toxic to plant life. Off-road recreational vehicles compact and severely harm fragile desert soils. In desert grasslands, overgrazing and trampling by cattle lead to a loss of native species and the predomination of weedy, undesirable species. Unlike the grassland ecosystems of the Great Plains, intermountain desert areas historically did not experience such large grazers as bison, and thus the grasses exhibit few adaptations to persist under grazing.[18] Therefore desert ecosystems are not well suited to conventional agriculture, even when irrigated.

Agriculture in the Desert's Image. What would an agriculture modeled after the native vegetation look like? It would have to make use of plant species that exemplified the various adaptive strategies shown by plants to the difficulties presented by the desert ecosystem. It would also place different types of plants across the array of soil types. Thus, some crops would be water-conserving succulents grown in dry areas. Others would be drought-hardy herbaceous or woody perennials with growth periods timed to use precipitation when it is available. Some would be legumes necessary for supplying nitrogen fertility to poor soils. Annual crops would be appropriate for alluvial fans, where natural nutrient and moisture subsidies occur.

In the southwestern United States, in fact, some indigenous cultures have successfully modeled their agriculture on different structural aspects of the desert ecosystem.[19] Such native peoples as the Papago and Cocopa cultivate some wild desert plants (for example, annuals, such as *Amaranthus palmeri*, and succulents, such as *Agave* species) and also some domesticated native desert plants (annuals, such as *Proboscidea parviflora*; perennial grasses, such as *Panicum son-*

orum; and legumes, such as *Phaseolus acutifolius*) in their traditional agriculture. These plants, already adapted to the desert ecosystem, are grown in harmony with local resources and seasonal availability of rainfall. Some of these food or fiber plants are annuals, some are succulents, some are grasses, and some are woody. Native diets are diverse and depend upon gathering from a great variety of available plants. In addition the native farmers allow such beneficial wild plants as mesquite trees to remain in their fields to provide nitrogen and gather deeply stored soil nutrients. Recognizing that periodic flood water loaded with floating detritus represents the most significant soil amendment for their fields, the Papago follow the distribution and emergence patterns of desert annuals by planting their annual crops in alluvial fans following favorable rains. They thereby avoid the ecological problems encountered by farmers who have moved in agriculture appropriate only for flood plains to the uplands or have artificially boosted the desert's productivity through excessive irrigation.

TEMPERATE DECIDUOUS FORESTS

The second example of an ecosystemlike agriculture is based on temperate deciduous forests. In North America, these forests cover the Atlantic Coast from eastern Canada to the southeastern United States and extend westward across the eastern third of the continent to the prairie bioregion. Thus, this ecosystem type occurs across a wide geographical and climatic range. Mean annual precipitation in temperate deciduous forest ecosystems ranges from 81 to 122 centimeters (32 to 48 inches) and is fairly evenly distributed throughout the year. Soils also vary widely, so much so that it is difficult to generalize about the soils of the eastern deciduous forest. These range from glacial deposits in the north, to soils derived from sedimentary and igneous rock, to rocky outcrops with shallow soils on uplands, and to relatively deep soils west of the Appalachian Mountains. How does this ecosystem function substainably across such a range of conditions?

Eastern deciduous forests are primarily wooded but have well-developed shrubby and herbaceous understories. These communities exhibit great structural diversity, with a high canopy, distinct subcanopy, shrub layers, and an herb layer at the forest floor. Typical total standing biomass—that is, trees, roots, understory, and herbs—may equal 9000 gm/m^2 (40 tons/acre). Of this, 10 percent to 30 percent may be in root biomass. Decomposition rates may be slow due to cool, shaded soil and high tannin content of fallen leaves. Litter and soil organic matter accumulations can be as high as 12,500 gm/m² (56 tons/acre).[20]

Although the longevity of trees makes the forest seem very stable at first glance, trees do fall periodically, creating large disturbances. These disturbances are critical to the maintenance of understory species and wildlife. Tree-fall gaps may represent as much as 10 percent of the total forest landscape, with approximately 1 percent of the total land area disturbed by tree falls each year.[21] In the gap created by a tree fall, light and precipitation suddenly begin to reach the ground surface. Understory plants increase in vigor, flower, and fruit before the canopy closes once again. Wildlife, such as birds and browsing ungulates, find these gaps to be meccas for feeding and cover.

Sustainable Features. Sustainable features of the temperate deciduous forest include fine, fibrous roots produced by woody understory plants that hold soil and retain nutrients, species that occupy different niches (for example, some species are restricted to forest floors whereas others produce leaves in the forest canopy; some understory plants grow and flower in the spring's brightness before the canopy leafs out; some plants produce evergreen leaves that allow photosynthesis when deciduous plants are dormant) and vertical and spatial diversity that favors natural pest control as well as habitat for many animal species. Nutrient cycling by the understory prevents both leaching and downslope loss of critical nutrients. Mycorrhizae further enhance the interconnectedness of forest species by directly linking decomposing litter on the soil surface with the active roots of trees and shrubs.

Agriculture Modeled on the Deciduous Forest. Prior to colonization and the clearing of North American deciduous forests by Europeans, Native Americans practiced agriculture in small parcels (8 to 80 hectares, or 20 to 200 acres) that were cleared and farmed for only a few years. Here, to supplement what they hunted, gathered, or acquired through trade, they planted maize, beans, squash, and tobacco. Because agriculture decreased the fertility of these forest soils, however, the sites were abandoned after a period of eight to ten years. Farming of a patch resumed after a ten-to-twenty-year fallow, during which the natural successional vegetation restored the fertility of the soil. Sustainability of Native American societies required them to be nomadic and to alter their diet throughout the year as they ate in accordance with what the ecosystem provided seasonally.[22] To a large extent, this indigenous style of shifting agriculture mimicked the natural forest's dynamism by creating small patches that were allowed to revert to forest.

In response to the destruction of hillsides in the eastern United States resulting from deforestation and row-cropping, J. Russell Smith believed that farming should fit the land.[23] He proposed nut- and fruit-bearing trees as the only fitting crops for such places. The agriculture he envisioned included a perennial understory that would fix nitrogen, hold the soil, and be suitable for grazing and haying. He hoped that an agriculture developed to fit moist hillsides would become an endless vista of crop-yielding trees. One example where such a structural model was practiced for a time was a hillside planting of honey locust with a Chinese bush clover (*Lespedeza sericea*) understory in Alabama. This system demonstrated effective soil protection, produced both hay and seed crops, supported grazing, involved low management costs, provided good weed control, and required relatively low labor inputs. The system yielded 5040 kilograms of hay per hectare per year (4500 pounds per acre per year) and averaged 3270 kilograms of honey locust nuts per hectare (2920 pounds per acre), with a peak of 9800 kilograms of nuts per hectare (8750 pounds per acre) in eight-year-old trees.[24] This model agricultural system embodied the sustainable features of the natural forest model: a tree crop as an overstory, a stable understory to protect the soil and retain nutrients, a biological nitrogen source, and a grazing or browsing animal component.

TROPICAL FORESTS

The third example features tropical forests as a model for agriculture. What are the critical characteristics of tropical forests that make them suited to their climate? What are the essential structural features that make them unique? What features would an agriculture modeled on a tropical forest require?

Many humid tropical forests are characterized by relatively constant temperature regimes but large seasonal differences in rainfall. Typically, these ecosystems experience both monsoon and dry seasons, with as much as a month passing without rainfall. Many tropical plants flower during the dry season, when the trees are leafless.

It is easier for pollinating insects and birds to find the flowers during this leafless period.

Soils of the humid tropics are characteristically low in organic matter but exhibit high rates of decomposition and nutrient uptake by plants. The combination of rugged topography and high rainfall results in appreciable soil erosion on slopes and subsequent deposition to the rich alluvial fields in valleys prized for intensive agriculture.

Tropical forests are about as far from monoculture as can be imagined. Exceptionally rich in plant species, Meso-American forests south of Mexico may boast 18,000 plant species. An average hectare of tropical forest contains some 100 different tree species (about 40 species per acre). In addition to the forest's great species and structural diversity, tropical ecosystems are perpetually changing. Frequent tree falls spawn local successional events that are essential for successful regeneration of up to half of all tropical tree species.[25] The average turnover time for the forests of La Selva, Costa Rica, has been calculated to be 118 years. Because such dynamics are important in maintaining these species-rich forests, it may be more accurate to think of tropical forest ecosystems as nonequilibrium communities than as equilibrium communities.

Sustainable Features. What, then, are the major features that lend sustainability to the tropical forest? A diversity of species—which rapidly fill available space both vertically and horizontally, and which quickly tap nutrients as they become available during the rapid decomposition process—is a critical feature of tropical forests. Since disturbances are so frequent, and tropical soils so fragile, the nutrient-retaining characteristics of the successional vegetation in gaps are critical to the sustainability of these ecosystems. This rapidly growing successional vegetation effectively protects exposed soil and takes up soil nutrients that would otherwise be leached or eroded.

Agriculture Modeled on the Tropical Forest. Agriculture modeled on the tropical forest must mimic its natural disturbance regime, shifting from site to site as succession produces new forest on disturbed sites. This successional agriculture would use groups of crops that mimic the vegetation characteristic of several intermediate (seral) stages and eventually give way to tree crops. An overall emphasis on tree crops and high species diversity is also essential.

Many traditional forms of tropical agriculture fit this model quite neatly. They rely on locally available energy and materials, as well as on the wisdom and information collected over the years by the people who have practiced these forms. Shifting or swidden agriculture mimics natural disturbances, with human felling and burning of small forest plots. The people then take advantage of the residual fertility characteristic of early succession and farm the plots. But nutrients are not available for long in tropical forests, and a plot may therefore be farmed only for one or two years.

Recall that in tropical forests most circulating phosphorus, potassium, calcium, and magnesium are found in plant tissues, not in the soil. And although the largest stocks of nitrogen and sulfur are in the soil, most occur in organic compounds that are not directly available to plants. During the process of felling and burning the forest, substantial amounts of nitrogen and sulfur are volatilized, and significant fractions of nitrogen, sulfur, phosphorus, potassium, calcium, and magnesium are lost through rapid mineralization followed by erosion and leaching. In particular, as much as 70 percent of the nitrogen and sulfur may be lost,[26] although some is recovered when the forest is reduced to ash.

Shifting agriculture takes advantage of this nutrient flush. When fertility has been depleted, and the weeds and tree sprouts begin to choke out the crops, the farmer abandons the field. During the fallow period the forest reclaims the land and the soil's meager fertility is rejuvenated. The length of this restoration period varies with climate and the agricultural pressure placed upon the land.

It is very difficult to sustain monocultures of annual crops on most tropical soils unless they are heavily subsidized with fertilizer and pesticides. High rates of nutrient loss via erosion and leaching dictate that the quantities of fertilizer added to maintain fertility exceed the amounts removed in harvest. Styles of agriculture developed for temperate climates frequently do not suit the tropics.

As noted above, the advance of natural successional vegetation following disturbance is critical to the sustainability of tropical forest ecosystems. The natural successional community that recolonizes newly deforested tropical soils is usually vigorous and productive and seldom appears, except on very poor sites, to suffer from critical shortages of nutrients. The rapid accumulation of biomass during succession means that substantial amounts of nutrients are removed from the soil, where they are potentially susceptible to leaching, and stored in plant tissues. With time, the internal recycling of nutrients that typifies more mature ecosystems increases as plants become less dependent upon new supplies of nutrients from the mineral soil. Eventually, during succession, plants appear with deep roots that are able to mine nutrients, an important aspect of the restoration process. Where the frequency of the cultivation/restoration cycle is low enough to accommodate the biological restoration powers of the ecosystem, shifting cultivation methods can endure successfully for centuries, if not millennia.

As reported in his article in a 1980 *Biotropica* supplement, Robert Hart designed a tropical successional agriculture for Central America that employs ecological analogs of successional vegetation ranging from colonizing annuals to overstory trees. In the first few years, the successional agroecosystem he proposed features such annual crops as maize and beans. The agroecosystem then undergoes a series of successional stages, which include cassava and banana, that eventually culminate in such tree crops as cacao and rubber. This seminal proposal directly related the agricultural viability of a cropping system to the structural, and thus functional, similarity between that cropping system and the natural ecosystem of a particular place.

In an interesting study, a team of scientists set out to mimic a natural successional tropical forest to see whether they could reduce the nutrient loss that occurs soon after clearing such a forest for agriculture.[27] This imitation of succession was an attempt to build an ecosystem that resembled the natural secondary growth of the area structurally and functionally, but using different species. In this study, the investigators replaced, as the plants appeared on a site, the naturally occurring species with morphologically similar ones (that is, annual for annual, herbaceous perennial for herbaceous perennial, tree for tree, vine for vine) not native to the site. In the treatment designed to imitate the species richness and general character of naturally occurring successional vegetation, the researchers removed naturally occurring colonists and in their place substituted morphologically similar species. Another treatment consisted of a temporal sequence of monocultures: two maize crops in the first year, followed by cassava (*Manihot esculenta*, a shrub that yields an edible tuber) for a year, then three and a half years of *Cordia alliodora*, a fast-growing timber tree. Two other treatments were natural successional vegetation and bare soil.

In general, the imitation of successional vegetation had the same structural and functional attributes as the diverse, naturally occurring successional vegetation. After four years, fine-root surface area, which is critical to site restoration following disturbance, was very similar between the native and the successional mimic. Fertility changes were markedly different among treatments, with depletion of essential cations fastest in bare soil and in monoculture treatments. Among monocultures, shorter-lived monocultures lost nutrients more rapidly than did longer-lived ones, due partially to the extensive fine-root mass produced by the *Cordia alliodora* trees. In contrast, the species-rich successional vegetation was effective in maintaining soil fertility, although general trends of soil nutrient decline were evident with time, presumably because of plant uptake. The successional vegetation and the imitation of succession had no species in common, yet they did not differ with respect to their impacts on soil fertility.[28]

This study showed that it is possible to achieve natural ecosystem function, in this case the fertility-maintaining characteristics of successional vegetation, by imitating the structure of that ecosystem with a mix of similar species. Thus, agroecosystems that incorporate such features of natural regrowth as extensive fine-root systems and perennial above-ground structure will reduce nutrient losses by storing large quantities of nutrients in biomass, by augmenting soil organic matter, and by protecting the soil from erosion.

Successful agroforestry systems in the tropics that involve mixed plantings of tree crops with herbaceous crop understories (for example, maize or cowpea) represent a more static, or less cyclical, equilibrium agriculture. Examples of this type of system are coffee and cacao plantations where shade trees are planted along with the crops. The leguminous trees are crucial to nutrient balances within the plantations. They can reduce nutrient leaching, fix nitrogen, and extract nutrients from deep soil.[29]

VEGETATION STRUCTURE AS THE KEY TO SUSTAINABLE FUNCTION

The three examples described above, based on deserts, temperate forests, and tropical forests, all have in common native ecosystems that did not fare well under conventional agriculture. In all three cases, though the physical conditions are vastly different, alternative systems of agriculture that mimic the native ecosystems appear to function sustainably. It was shown that, whereas irrigating the desert and growing monocultures of annuals renders the soil too salty to farm and uses up vast quantities of fresh water, farming in nature's image—with annuals used only in appropriately moist areas and seasons of the desert—is indefinitely sustainable. Whereas farming the steep slopes of eastern North America destroyed soil in less than two centuries, mimicking the native forest, even in a simplified manner, allows these slopes to be farmed for tree crops. Whereas importing industrial-style monoculture agriculture to the tropics destroys soil and requires tremendous inputs of nutrients in some cases, mimicking the natural successional process, even with non-native species, permits nutrient conservation and soil preservation and restoration. In all three cases the structure of the native vegetation holds the key to a sustainable agriculture for that region.

A SYSTEM FOR THE NORTH AMERICAN GRANARY MODELED ON THE PRAIRIE

This last conclusion can be applied to one more ecosystem. Over a short 100 years of applying monoculture farming to the native prairies of North America, 50 percent of the topsoil's latent productivity has slipped away. In contrast, the native prairie built and maintained soil and supported large populations of grazers. Perhaps before the remaining 50 percent of latent soil productivity is used up, a model for the future for North America's breadbasket, a model that mimics the native vegetation, should be considered.

THE PRAIRIE ARCHETYPE

As shown in the previous sections, successful agricultural mimics of various ecosystems can be designed and established to produce vegetable, tree nut, and grain crops. Since people are primarily grass and, secondarily, legume seed eaters, however, it is most useful to look to an ecosystem that features grasses as the dominant plant type for the best model for a sustainable grain agriculture. The prairie, therefore, is an appropriate natural model for grass seed agriculture on the Great Plains. The North American prairie is characterized by wide open landscapes of primarily herbaceous vegetation. In the past, as farmers replaced arrays of native grasses and forbs with such alien species as wheat, sorghum, and soybeans introduced from other continents, North American landscapes were modified to accommodate the biological requirements of these new plants. Yet the remnants of intact prairie serve as excellent examples of inherently sustainable biotic communities in which complex webs of interdependent plants, animals, and microbes garner, retain, and efficiently recycle critical nutrients. Prairie plants have adapted successfully to the climate, insects, and diseases of the highly variable Great Plains.

Climate, Fire, and Large Grazers. Climatic extremes characterize, indeed maintain, the North American prairie. Annual precipitation ranges from a maximum of 100 centimeters (40 inches) per year in eastern Texas to 25 centimeters (10 inches) per year at the prairie's western border. About 75 percent of the precipitation falls during the growing season. Like all grassland ecosystems, the prairie has a period of seasonal drought, and in this case it is a summer drought. Temperature ranges are also great. In the tallgrass prairie region of Kansas, annual temperature may range from $-40°C$ to $45°C$ ($-40°F$ to $115°F$), and even diurnal temperature may range $40°C$ ($70°F$) within a twenty-four-hour period.

The prairie region's climate extremes also contribute to the great store of fertility in prairie soils, which are among the richest soils in the world. Seasonal drought cycles, involving warm, moist springs favorable to luxuriant grass growth followed by dry summers, have led over the millennia to an accumulation of soil organic matter. It is this store of organic matter that has made the highly productive U.S. grain belt possible. We have essentially been mining the fertility built up over millennia by the native ecosystem. In prairies, soil is built through the turnover of roots. In some prairies, 30 percent to 40 percent of roots may turn over each summer. Indeed, most of the prairie's biomass occurs underground as roots, rhizomes, and crowns. These underground parts represent from 60 percent to 75 percent of total plant biomass in the prairie.

Besides climate, two other components are critical to the development and maintenance of the prairie: fire and grazers. Historically, fire was frequent because conditions were often ripe. The large, open prairie landscapes had few natural firebreaks. Extensive tinder accumulated each year in the form of dry plant litter, and summer storms with frequent lightning strikes provided the sparks to set off fires. Adaptations possessed by prairie plants to withstand fire include underground buds (meristems) for new growth and increased vigor following a burn. On the other hand, fire damages shrubs, trees, and actively growing cool-season plants. Largely migratory herds of bison were the major grazers of the prairie ecosystem. Adaptations of prairie species to endure grazing include underground meristems and a rhizomatous growth form (spreading by underground stems) that resists trampling damage.

Flexible Structure. What are the important structural features of the prairie that accommodate fluctuating climate, fire, and large grazers? The first aspect of native prairie that strikes the observer who looks closely is the great diversity of plant species growing side by side. In Kansas tallgrass prairie, showy blue- and lavender-flowered legumes are intermixed with tall, yellow composites and dense grasses. Observing the soil surface closely, one notes barely any room among shoots for any more plants. A quarter-meter-square sampling area may contain up to 25 different plant species (equal to 9 species per square foot). This diversity turns out to be one of the crucial features that lends the prairie sustainability.

Prairie vegetation is a grass-dominated mixture consisting primarily of perennial C_3 and C_4 grasses, legumes, and composites. These components form patterns that change from place to place and over time. On a broad spatial scale, the ratios of different types of plants vary over the prairie region in response to climatic patterns. Cool-season grasses predominate in the northern prairies of Canada, whereas warm-season grasses predominate in the southern prairies of Kansas, Oklahoma, and Texas. The species composition varies from east to west also. The tallgrass prairie region, which formerly constituted the eastern third of the Great Plains, is dominated by big bluestem (*Andropogon gerardii*), Indiangrass (*Sorghastrum nutans*), and switchgrass (*Panicum virgatum*). In the Dakotas, wheatgrass (*Agropyron* species) and needlegrass (*Stipa* species) are additional major components. In the shortgrass prairies farther west, the grasses are mainly gramas (*Bouteloua* species) and buffalograss (*Buchloë dactyloides*).

On a local scale, the proportions of different plant groups vary across soil types. Species diversity is generally higher on poorer soils, whereas rich soils support a vegetation that approaches single-species stands. Not surprisingly, more legumes tend to occur on poorer soils.

Moreover, prairie communities are dynamic in time, changing in response to long-term climatic cycles. Species composition on any one site changes in response to the vagaries of climate or disturbance. During this century the North American prairie has experienced dry periods approximately every twenty years. During the drought of the 1930s, western wheatgrass (*Agropyron smithii*) supplanted tallgrasses on much of the eastern Kansas prairie, and many shortgrass species moved eastward. When normal precipitation resumed in the 1940s, the prairies reverted to their predrought species compositions. Long-term studies of this sort have shown that the prairie community can rotate over a period of many years. This allows a vegetation complex to endure while adjusting to cyclic climatic fluctuation. This helps explain the renowned resilience of prairies.

Species Complementariness. In large part, the resilience of prairies is due to a diversity of plants with complementary niches. One important dimension of complementariness is how different species cope with the extremes of the Great Plains climate. Some species are adapted to endure the extremes, but others are adapted to avoid them. Grasses with C_4 photosynthesis and drought-hardy forbs are able to withstand the hot, dry conditions of summer. Others, the cool-season grasses and forbs, persist by avoidance, by growing and setting seed before the onset of summer drought. The differences in timing of resource needs in these two groups means that the prairie uses available resources more completely. Resource use is spread across the entire growing season, and competition for such things as soil nutrients, moisture, and pollinators is less than in a field of plants all at the same growth stage.

Under the ground, plants further differentiate their roles and subdivide the environment. One plant may produce a deep taproot, while its neighbor produces shallow, fine roots. These differently shaped roots not only reduce direct competition among neighbors for rooting space, but also correspond to different ways of obtaining nutrients. Many thick-rooted prairie forbs and C_4 grasses are dependent upon mycorrhizae. For example, big bluestem cannot efficiently obtain phosphate without mycorrhizae. On the other hand, because fungi are less metabolically active at low soil temperatures, many cool-season grasses are not mycorrhizal, but have fine roots instead. Thus, mycorrhizae can potentially influence the species composition of prairie communities by influencing the relative competitive abilities of different species and thus their distribution.[30]

The complementariness of the different prairie species' niches allows many species to coexist, so that when disturbances occur or climate changes, a pool of species that can take advantage of the change is always available. Resilience, provided by species complementariness, is one substainable feature of the prairie.

Tight nutrient cycling is another sustainable feature. Because most nutrients are tied up in living biomass and soil organic matter, they are not vulnerable to leaching or erosion loss. Nutrients are cycled seasonally within plants, stored in organic matter, and quickly taken up once mineralized by decomposers.

The activities of plants are greatly influenced by the other members of the prairie ecosystem. For example, below-ground/aboveground relationships are coupled by the activities of herbivores and the decomposer community. Immediate levels of grazing by ungulates can increase the degree of root turnover and thereby increase densities of root-feeding arthropods,[31] which can in turn raise rates of nutrient cycling. This effect could be detrimental to plant growth at higher levels of grazing.

Thus, the features that enable the prairie to endure in a region of climatic extremes are a diversity of species with a variety of strategies for coping with the climate; a component of legumes to provide nitrogen where needed; tight nutrient cycling; species that use different parts of the soil at different times of year and thus effectively capture available water and nutrients; tightly coupled plant and animal interactions; and perennial cover to protect the soil.

AN AGRICULTURAL MIMIC OF THE PRAIRIE

How can an agriculture be structured to mimic the prairie ecosystem? As opposed to the monocultures of annual crops—namely, corn, wheat, sorghum, and soybeans—that presently occupy most of our prairie soils, an agroecosystem modeled on the prairie would be dominated by mixtures of perennial grasses, legumes, and composites as seed crops, whose species composition would vary across soil type and climate. Polycultures of herbaceous perennial seed crops, based on the prairie community as a model, would be composed of plants that differ in seasonal nutrient use and would thereby play complementary and facilitating roles in the field. Promising candidates for such a perennial agriculture include the C_3 grasses wildrye, or leymus (*Leymus racemosus*) and intermediate wheatgrass (*Agropy-*

ron intermedium), the C$_4$ eastern gamagrass (*Tripsacum dactyloides*), the legume Illinois bundleflower (*Desmanthus illinoensis*), and the Maximilian sunflower (*Helianthus maximilianii*), a composite.

Potentially, such an agriculture would tap many of the sustainable features of the prairie. The use of perennial species would tap the soil-retaining, soil-building aspects of the prairie. The legume component would help maintain an internal fertility supply. The variety of climatic adaptations and seasonal variation in growth and reproduction would lend resiliency and promote efficient use of available resources. A diversity of crop species, including some native species, would allow development of some natural checks and balances on herbivores, diseases, and weeds.

Some of the broader implications of a grain agriculture based on perennial crops are conservation of natural resources, reform of the current economic system to be more ecologically based (that is, nature as a model for economics), and a society that engages nature on nature's terms. An ecology-based economic system would take into account available ecological capital, feature steady states of resource cycling, and efficiency of energy transfer and would rely on available solar or biological energy. Some of these broader implications are explored further in Chapter 6.

By combining these expected sustainable features with the broader implications for society and the environment, a list of potential agricultural benefits of perennial polyculture can be formulated: (1) reduced or eliminated soil erosion, (2) efficient use of land area and soil nutrients, (3) increased water-use efficiency, (4) reduced reliance on industrially produced nitrogen fertilizer, (5) decreases in pest and disease epidemics, (6) effective chemical-free weed management, (7) reduced energy use in tillage, (8) reduced chemical contamination of soil and water, and (9) insurance against complete crop failures. What are the aspects of the model that will promote these agricultural benefits?

Ecological Benefits of Perennial Polycultures. First, by using perennial crops instead of annuals, the model taps the soil- and nutrient-conserving properties of the prairie. Perennial crops can provide constant vegetation cover and root mass to promote soil accumulation and nutrient retention and to reduce erosion. The continual seasonal root turnover of perennials builds soil organic matter, improves structure and porosity, and enhances nitrogen fixation by free-living microbes. Roots are generally the most important source of soil organic matter. For example, in forests, roots and mycorrhizae may contribute up to five times as much organic matter to soil as do fallen leaves and branches. When compared to annual crops, herbaceous perennial crops allow more soil carbon and nitrogen (if legumes) to accumulate, prevent leaching, and promote greater activity of the decomposer community.[32]

As pointed out by Cory W. Berish and John J. Ewel in "Root Development in Simple and Complex Tropical Successional Ecosystems," an agronomically important aspect of perennial plants is that organic matter from root turnover is added to the soil nearly continuously, whereas in annual crop fields most organic matter is returned to the soil all at once, following harvest. If decomposition and mineralization of plant matter occur at the time of reduced plant growth, soil nutrients can be lost via leaching or erosion. Turnover of roots and soil organisms, without plowing and the accompanying oxidation of soil carbon by decomposers, will result in added organic matter. Thus, soil structure should improve and a healthy soil community should be maintained.

These soil-building features of perennial crops are excellent for reducing soil erosion. Water-holding capacity increases with increased organic matter content. This both reduces run-off and consequent erosion and conserves soil moisture. Root-held soil and constant cover can eliminate erosion even on sloping ground.[33]

The second aspect of the model that promotes agricultural benefits is the combination of compatible perennial crops into mixtures. Perennial polycultures would tap all the advantages of intercropping systems. Space and nutrient resources would be used relatively efficiently due to the complementary niches among crop species.

Third, crops derived from species adapted to seasonally dry grassland prairie environments should display water-use patterns that are synchronized with the seasonal availability of precipitation. Thus, water use by these crops should be inherently efficient. Alternatively, cool-season crops will flower and fruit early in the growing season, thus avoiding much of summer drought. Moreover, because surface run-off is reduced by a leafy layer that breaks up rain droplets, precipitation is more likely to soak into the soil. Less run-off from agricultural fields means more effective precipitation as well as the reduced likelihood of siltation and eutrophication of nearby surface waters.

Fourth, perennial polycultures would feature nitrogen from legumes. Legumes are a critical part of many intercrop and rotational schemes. Direct transfer of nitrogen from living legumes to grasses has been demonstrated experimentally. Increased reliance on biological fertility reduces the need for purchased inputs.

Fifth, a decrease in pest and disease epidemics would be expected through the benefits of species diversity in the field. This diversity would both protect host plants from pests as well as encourage and maintain populations of beneficial predators.

Sixth, weed management would occur through the perennial, shading canopy of the new crops. If one or more crops were allelopathic (for example, the Maximilian sunflower) against weed seedlings, then additional weed control could be expected.

Seventh, the amount of fossil energy used for traction in seed bed preparation and cultivation, application of synthetic chemicals, and irrigation would decrease in perennial polyculture agriculture. These energy uses would decline because yearly seed bed preparation is unnecessary with perennials, the need for cultivation is reduced due to the continual canopy cover, and fertility and water balance are maintained as discussed above. Additional energy would be saved at the level of manufacture of equipment and synthesis of fertilizers and biocides. Renewable solar-biological energy would instead supply a higher percentage of agricultural energy needs, and biological factors would help manage undesirable organisms.

Eighth, greater reliance on legume nitrogen and biological control of pests and weeds should lead inevitably to reduced chemical contamination of soil and water. Inputs of various chemical biocides would decline as effective protection is provided by genetic diversity within and among species, encouragement of predators and parasitoids, management schemes that minimize the factors favoring plant disease, and weed control through shading and allelopathy. Inherent protection is obtained by emphasizing crops adapted to avoid the ravages of endemic insects and pathogens.

Ninth, diverse agriculture, mixtures of different species as well as genetic variants within, avoids the loss of genetic diversity that has plagued conventional agriculture in recent decades (see Chapter 1). Recall that genetic variability is a characteristic of wild plant populations. As discussed earlier, diversity in both wild communities and in agricultural fields confers population- and community-level resistance to disruptive change. Wild areas maintained in conjunction with agroecosystems could further enhance the genetic diversity of the landscape and serve two important additional functions: to provide refuges for beneficial predators and to remain as ecological standards for agricultural designs. Using different varieties adapted to local conditions in different habitats would maintain genetic diversity within species across the ranges in which they are grown. A selection program at the community level would capture and maintain the benefits of genetic diversity, interspecific compatibility, resistance, attraction of beneficial insects, and allelopathy.

Origins of the Prairie Model for Agriculture. A perennial polyculture of grain crops is really a domesticated prairie. It is admittedly a radical departure from current agriculture. It involves new crop species that are selected under much more complex regimes than are used for crops destined for monoculture. It involves new designs for crop mixtures and new forms of attention from farmers. It requires new time lines for financing and calculating economic returns. It involves more local fine-tuning than is now used. But it is not totally unexplored. First proposed by Wes Jackson in his 1980 book *New Roots for Agriculture,* and updated in 1985, the vision of perennial polycultures of grain crops has received widespread attention in the press and from researchers interested in alternative agriculture. But research into perennial polycultures has not yet found a home in traditional agricultural research institutions. This is not surprising, given the unique mix of research expertise required and the necessity of a multidisciplinary approach, not to mention the difficulty that traditional research institutions have with radical departures from the status quo. Therefore, The Land Institute, a nonprofit teaching and research institution founded in 1976 by Wes and Dana Jackson in Salina, Kansas, remains the center for this research.

At The Land Institute, researchers[34] are practicing the type of multidisciplinary research many feel is needed to move toward a sustainable agriculture. The focus there blends ecology and agriculture to develop mixtures of perennial grain crops as analogs of—that is, crops that function similarly to—the native grasslands of the Great Plains. Thus, the goal of their work is to discover the ecological principles that promote sustainability within such agronomic mixtures. The research on perennial polycultures, described in detail in Chapter 5, represents one of the very few instances in which an ecologist is collaborating with a plant breeder on the development of a new form of agriculture. To this end, The Land Institute serves as an important liaison between the alternative agriculture movement and land grant institutions. A cadre of university scientists has formally endorsed this work, and some of the scientific objectives are beginning to be adopted within the USDA (for example, the LISA—Low-Input Sustainable Agriculture—program) and research universities. The principles of sustainable agriculture are likely to be transferable and applicable across a wide array of agroecosystems.

Building on Other Research in Alternative Agriculture. Although perennial polycultures as envisioned would be radically different from current U.S. agriculture, their development need not be a matter of starting from scratch. Research on a prairie analog for agriculture builds on and relates to the established work in conservation tillage, biological control, intercropping studies, and ecological studies of natural ecosystems and communities discussed earlier in this chapter.

Since tillage will be much reduced in perennial polycultures, where crops need not be established every year, the knowledge already gained from conservation tillage research will be a valuable aid. But prairielike agriculture takes reduced tillage several steps further. The added feature of a perennial canopy or an allelopathic component addresses the weed-control problems of conventional conservation tillage. Continual input and organic matter supplied by perennial roots can build soil organic matter more rapidly, provide a more usable and constant nutrient supply, and protect soil better from erosion than annual crops can, even when grown with no-till methods.

Those who design an agriculture modeled after the prairie can also learn much from IPM about biological and cultural controls. But again, the model takes a giant step beyond IPM by incorporating structural and genetic diversity components of natural pest control and long-term stability of the soil community. Crop diversity provides both genetic and physical interference for pests and promotes natural enemies of pests.

From intercropping work, researchers working on perennial polycultures can borrow knowledge of the techniques for establishment and harvest and how physical arrangements of component species affect yields. This research will build on intercropping theories by going beyond two-component systems and by using perennials instead of annuals. It may also incorporate succession, with the dominant components changing over a several-year period to mimic natural succession. And finally, it will use crop plants with some evolutionary or selection history in common. In other words, it will use natives of the same ecosystem as crops, or at least select crops under polyculture conditions, so that part of their selective history is in common. This will ensure compatibility of the component species.

From studies of prairie ecology, perennial polyculture research will build on knowledge of the factors that provide sustainability to healthy ecosystems and of the dynamics of natural plant populations and communities. These are the most basic building blocks for a prairielike agriculture. Knowledge of prairie function provides the only valid benchmark against which to compare the sustainability of agriculture. An agriculture derived from the prairie archetype, then, must be founded on features of ecosystem function discovered in studies of the prairie community. Knowledge of such functional and structural aspects as differences among species in growth period, nutrient use, and water requirements is directly applicable to the design of a domesticated prairie. To this end, ongoing research must compare patterns of nutrient cycling, ecological succession, long-term yield stability, and biological management of detrimental organisms between native prairie and perennial polyculture.

But the polyculture research must go beyond basic ecology and apply these ecological principles to create agronomically successful systems. It must also look at long-term patterns of seed yield to try to stabilize overall crop yields over a number of growing seasons. Likewise, what is discovered about the relative importance of competition and facilitation between species, nutrient cycling, and biological management of weeds, herbivores, and diseases must be applied to the agronomic problems of weed control, sustained yield, fertility management, and pest control. Similarly, knowledge of how the prairie responds to changes in climate and landscape will allow fine-tuning of species composition within or between fields across heterogeneous landscapes and regions.

Thus, by seeking a pattern that works by creating a structural mimic, researchers would simultaneously incorporate into agriculture the best of the process models. The biological control of pests as practiced in IPM would be incorporated via crop diversity and the promotion of natural enemies. Overyielding and crop stability would be provided through intercropping perennials. All the sustainable advantages of no-tillage agriculture would accrue to fields of perennial crops.

New Horizons for Sustainable Agriculture Research. What is new about perennial polyculture research as compared to other agricultural research, both conventional and alternative? First it incorporates the ecological perspective, which is holistic, complex, and long-term. By proceeding at the ecosystem level, it assumes that the successful system will not be explainable by an in-depth study of each of its parts. In addition, it involves complex new breeding problems. For example, very little ecological or genetic information on crop varieties appropriate for polyculture designs has been gathered, and plant breeders are only beginning to approach the problem of how to apply such knowledge to designing viable intercropping systems.

Much research has been conducted in selecting optimal densities and varieties for monoculture, but not much is known about the appropriate management of mixed-species plantings. For example, determining a plant species' suitability to polyculture can take place only within the context of its community; its agronomic characteristics in mixture may not be predictable from its behavior within monoculture stands. Breeding for perennial polyculture means that, because plants bred in monoculture may behave differently when grown in mixture, all selection designs must be complex. Sometimes a variety's performance in monoculture is a good predictor of its performance in polyculture, but other times its performance may vary when it is grown with different species or in different arrangements.[35] This becomes even more difficult when varieties show different responses depending on the type of cropping system in which they are grown. In these cases, all possible combinations of varieties must be compared. Such varietal differences as plant stature (erect versus prostrate) or phenology (days to flowering) are likely to lead to different behaviors within different cropping systems.

Another unique feature of research into the prairie model for agriculture is the long-term aspect of the ecological perspective. The crops to be grown in perennial polyculture may not yet exist. Thus, an inevitably long-term domestication process is involved. Because we may not see great progress for a few years, such an endeavor requires great patience on the parts of both scientists and funding agencies.

A long-term perspective is also required simply to think in terms of longer yield cycles between plantings. Because perennial crops will be required to maintain themselves in the field for a period of years, instead of months as with annual crops, it is necessary to incorporate multiyear demographic patterns of growth and seed yield into crop development research. For example, will the species yield its first year or remain vegetative? Will yields increase or decrease over time? Will yields be predictable year to year, or will they oscillate wildly? How will interactions among crop species in polyculture change over years? Unlike the grower of annual crops, who has some flexibility in changing crops or modifying field conditions after each growing season, the grower of perennial polyculture crops will have to project over a number of possibly very different growing seasons.

Thus, to create a domesticated prairie, much new research ground must be broken, but hopefully, in the process, the broken ground of the prairie will be healed.

5

The Feasibility of a Prairielike Agriculture

CHAPTER 4 DISCUSSED THE potential benefits of an agriculture, modeled on native prairie, composed of mixtures of perennial seed crops for human consumption. This agriculture would, it is hoped, yield sufficiently, provide much of its own fertility, and manage adequately troublesome weeds, insects, and plant diseases. The prairie is a structural and functional model for perennial polyculture research. Of course, the analogy is not perfect, as the prairie produces only small amounts of seed. By mimicking the prairie's structure—that is, herbaceous perennial plants growing in diverse, complementary mixtures—some of the sustainable features of the prairie community would be incorporated into agriculture. These features include soil building through the turnover of roots, reduced soil erosion, biological nitrogen fixation, efficient use of soil resources via minimal overlap among plant species, and stability and resilience of crop fields in the face of climatic extremes.

This chapter explores the feasibility of designing an agriculture that incorporates the principles of prairie ecology by mimicking the structure of the prairie. The purpose here is to support the scientific feasibility of the proposal with experimental data related to four critical subjects: sufficiency of seed yields from perennials, yield advantages of polycultures over monocultures, soil fertility, and relationships among crops, weeds, pests, and diseases in polyculture and monoculture.

The chapter begins by introducing four basic questions that guide much of The Land Institute's research toward a sustainable agriculture. It then recounts studies from the literature that support the feasibility of developing successful seed-producing perennial polycultures. It continues with the results of studies at The Land Institute specifically designed for the purpose of developing this form of agriculture. As a way of inviting others to join in this work, the chapter concludes with a list of some of the broader research in pure and applied sciences (plant ecology, plant breeding, soil science, population genetics, entomology, and animal science) that is needed to develop workable perennial polycultures.

FOUR BASIC QUESTIONS

The work at The Land Institute thus far has involved a synthesis among plant breeding, ecology, and plant pathology by staff scientists and formal and informal collaborations with agronomists, plant breeders, soil scientists, and entomologists at universities. Ongoing research encompasses four primary questions critical to the development of viable perennial polycultures.[1]

QUESTION ONE: SEED YIELDS OF PERENNIALS

First, can an herbaceous perennial seed crop yield as well as an annual crop? The last chapter made the case that perennial crops benefit the soil by building soil organic matter and eliminating erosion on sloping ground. While these benefits may be generally recognized, some scientists argue that because perennial plants must divert some photosynthate to below-ground storage to survive the winter, it may be difficult to breed these plants to produce as much seed as an annual crop, which dies after reproducing. Yet perennial species differ greatly in both relative and absolute amounts of energy devoted to seed production. Central to the thesis of this book is the authors' conviction that a fast-growing, well-adapted species could yield adequate seed while retaining the ability to overwinter. Because this is a pivotal issue in the development of perennial seed crops, the next section explores this possibility in detail.

Transcending this consideration is the question of whether sufficient genetic potential exists to domesticate and breed for stable, high yields in herbaceous perennial species. Recall that the development of perennial seed crop polycultures will require the domestication of new perennial crops from wild species into varieties that perform well in multiple cropping systems. The degree of crop improvement possible depends upon the amount of existing genetic variability among wild plant populations. To assess this variation requires collecting samples across the geographic ranges of candidate plant species and then evaluating these samples, or accessions, within a common garden.

QUESTION TWO: OVERYIELDING IN PERENNIAL POLYCULTURE

Second, can a polyculture of perennial seed crops outyield the same crops grown in monoculture? Overyielding can occur, for example, when interspecific competition in a plant community is less intense than intraspecific competition. Overyielding may also be found when one species enhances the growth of another. Thus, through differences among species in resource use, timing of demand, or rooting zone, or through direct beneficial relationships among species, multispecies fields typically yield more per unit area than do equivalent areas planted to monoculture. This is largely untested with the crops advocated here, however.

QUESTION THREE: INTERNAL SUPPLY OF FERTILITY

Third, can a perennial polyculture provide much of its own fertility? Specifically, this question asks to what extent such internal factors as nitrogen fixation by legumes and phosphate accumulation via mycorrhizae can compensate for nutrients removed in harvested seed. To answer this requires careful documentation of nutrient pools in the soil, rates of uptake by plants, nutrient content of harvested seed, and the capacity of crop plants and associated symbionts to enrich the soil over time.

QUESTION FOUR: MANAGEMENT OF WEEDS, PESTS, AND PLANT PATHOGENS

Fourth, can a perennial mixture successfully manage weeds, harmful insects, and plant pathogens with little or no human intervention? Weeds may be reduced either allelopathically or via continuous shading of the soil surface by perennial crops. Insect pests often can be managed through a combination of predator attraction, chemical or physical inhibition of insects' location of host plants, and trap-cropping. Disease control may be achieved by a combination of resistant lines and species diversity in the field. For many annual crops, a mixture of species, and of genotypes within species, can reduce the incidence and spread of some plant diseases, but less is known about this effect in mixtures of perennials.

RELATED STUDIES

As concluded at the end of Chapter 4, to a large extent work at The Land Institute builds on a substantial body of research in ecology, IPM, conservation tillage, and intercropping. It also builds on related work in perennial seed crop development. This section draws on research from sources other than The Land Institute to support the feasibility of perennial polyculture.

SEED YIELDS IN PERENNIAL PLANTS

Much of the research in perennial seed crops (Question One) has logically focused on perennial grasses, as the bulk of human nutrition is derived from grass seed either directly or indirectly through our consumption of grain-fed livestock and livestock products.

Early agriculturalists took the easier route in developing grain crops from annual rather than perennial grasses for important evolutionary reasons. For one, the annuals were likely already well suited to the sort of disturbance that annual cultivation creates. The closest wild relatives of many domestic grain crops are aggressive annual colonizers of disturbed habitats. In fact, some of our present-day grain crops (rye and oats, for example) were once weedy invaders of grain fields.[2] Presumably, these species were domesticated originally because they possessed certain characteristics that made their seed inviting as food. For example, they germinated quickly, completed their life cycles quickly, and produced relatively large seeds. As early agriculturalists purposefully sowed and harvested these plants, they consciously or unconsciously selected for such characteristics as reduced shattering, uniform maturation time, large number of seeds per stalk, naked and easily threshed seeds, and large seed size.

On the other hand, many perennial plants, though producing very nutritious seeds, lack the other characteristics that would lend them immediate usefulness as crops. Despite this drawback, however, many wild perennial grasses have provided food for humans in the past. For example, gathered seed from such perennials as wheatgrass (*Agropyron* species), bentgrass (*Agrostis* species), wildrye (*Elymus* and *Leymus* species), Junegrass (*Koeleria pyramidata*), Indian ricegrass (*Oryzopsis hymenoides*), and saltgrass (*Distichlis palmeri*) was important to the historical diets of many Native Americans.[3] Similarly, seed collected from perennial grasses was important in the past to people of northern Africa.[4]

Breeding Perennial Seed Crops. Two distinct approaches are possible in the modern development of perennial grains. One approach starts with annual grain crops and attempts to turn them into perennials via wide hybridization with perennial relatives. The rationale for this approach is that annual crops have proven themselves to have already undergone selection and domestication over the last several thousand years. Hence, they already show such agronomically favorable characteristics as edibility, high yield, large seed size, ease of threshing, synchronous maturity, and resistance to shattering and lodging (falling over). Wild perennials, on the other hand, rarely display these traits.

Unforunately, to date, most wide crosses (crosses between two different species) between annual grain crops and their wild perennial relatives have resulted in hybrids that are weak or sterile. For example, in crosses involving wheat or barley with wildrye or wheatgrass species, Soviet plant breeders usually obtained specimens that were weak, sterile, had poor seed set, and were annual. In another example, crosses between corn (*Zea mays*) and the perennial *Zea diploperennis* failed to produce winter-hardy, perennial offspring. Some hybrids between rye (*Secale cereale*) and the perennial *Secale montanum* yielded well in the first year (2200 to 2400 kilograms per hectare, or 1960 to 2140 pounds per acre), but yields declined afterward.[5]

The second approach starts with wild perennial plants and attempts to turn them into domestic seed crops. The theory here is that the potential crop already has the desired perennial trait but lacks most of the necessary agronomic traits. Characteristics that make a perennial species well adapted to its natural habitat are generally not the characteristics that make it a good grain crop. In other words, in the wild, seeds are small, do not thresh easily from the hull, do not mature simultaneously, and tend to disperse or shatter from the head easily. As noted earlier, most wild, herbaceous perennials emphasize vegetative growth, and seed set and yield are typically low. The difficulty facing the plant breeder who takes this approach is to overcome these undesirable characteristics while maintaining perennialism and good nutritional qualities of seed. For example, intermediate wheatgrass (*Agropyron intermedium*) typically produces low yields, but the grain is of good quality. Saltgrass, a perennial halophyte of salt flats and marshes, displays very low yield (about 200 kilograms per hectare, or 179 pounds per acre) but shows great promise as a drought- and salt-tolerant grain.[6]

With either approach, obtaining high seed yield from perennials is more complex than it is with annuals. Not only must the yield be high in one year, but to be profitable, yield must be maintained at sufficiently high levels for several years after the crop is established.

It is clear that these two challenges are related, since a perennial must divert some resources away from seed production each year to maintain overwintering structures. In annuals all of a plant's available photosynthate goes to the seed, and vegetative parts die at the end of the growing season. In wild perennials, on the other hand, survival depends on maintaining the perennating structures (overwintering crown, rhizomes, and root), and recruitment from seed is relatively unimportant to fitness. Consequently, the percentage of plant resources available for seed production is far less than in annuals. Thus the plant breeder must pay attention to both seed yield and maintenance of overwintering ability.

The second challenge, maintaining consistency of yield over time, may be addressed by careful management in addition to breeding. Under cultivation, yields of many herbaceous perennials typically decline after the first reproductive year, but some management techniques can help slow this decline. Therefore, developing higher yield in perennial plants involves both the selection of consistently high-seed-yielding lines and devising cultural techniques to maintain high yield.

Plant breeders generally recognize four direct components of seed yield: percentage seed set, seed size or weight, size of inflorescence (or number of seeds per stalk), and number of reproductive shoots or tillers per plant. Factors that can influence seed yield indirectly are synchrony of flowering, which affects interplant fertilization, resistance to lodging, resistance to shattering, and resistance to disease and insect pests.

Seed Yields of Some Perennials. Whichever approach is taken in domestication, seed yields of any successful perennial crops must be at some acceptable level to make the perennial polyculture model compelling. For comparison, the bench-mark yield for Kansas wheat is 30 bushels per acre, with each bushel weighing 60 pounds. This translates to a yield of 1800 pounds per acre, or about 2000 kilograms per hectare. Of course, wheat yields are higher in other regions, and in particularly favorable years North American corn yields run closer to 9000 or 10,000 kg/ha (8000 to 9000 lb/ac). A few records for both woody and herbaceous plants approach these bench-mark yields. Such important nut trees as the chestnuts (*Castanea* species), the walnuts (*Juglans* species), the hickories and pecan (*Carya* species), and the oaks (*Quercus* species) typically yield from 1100 to 1800 kg/ha (980 to 1600 lb/ac).[7] The desert shrub jojoba (*Simmondsia chinensis*) is being developed as a water-efficient oil-seed crop for arid areas. In stands monitored over several years, yields of jojoba have steadily increased over time in both no-input and fertilized, irrigated plots. In fact, in some fertilized and irrigated plots, yields have exceeded 2000 kg/ha (1800 lb/ac); low-input plots have performed less well.[8]

Most of the seed yield studies for unimproved (wild) herbaceous plants are for grasses and legumes. At the Rodale Research Center in Kutztown, Pennsylvania, intermediate wheatgrass has produced as much as 566 kg/ha (505 lb/ac), but first-year yield averaged across experiments and years was only 362 kg/ha (323 lb/acre).[9] Although most crosses were only weakly perennial, grain yields of various winter-hardy hybrids between perennial rye and rye ranged from 620 to 1555 kg/ha (554 to 1388 lb/ac) in the first year and from 575 to 910 kg/ha (513 to 812 lb/ac) in the second year.[10] Seed yields of wild perennial plants range from low to relatively high. For example, stands of switchgrass (*Panicum virgatum*) grown without fertilizer produced 652 kg/ha (582 l6/ac) when irrigated and 232 kg/ha (207 lb/ac) when not irrigated.[11] On the other hand, buffalograss (*Buchloë dactyloides*) produced in one year 1811 kg/ha (1617 lb/ac) under irrigation, and showed a four-year average yield of 1005 kg/ha (897 lb/ac).[12] Dryland, unfertilized grass yields in the Great Plains range from 500 to 1000 kg/ha (450 to 890 lb/ac), whereas fertilized and irrigated yields of perennial grass seed can approach 2000 kg/ha (1800 lb/ac).[13] Similarly, irrigated and fertilized perennial legumes can yield between 500 and 1100 kg/ha (450 and 980 lb/ac).[14]

Low yield in a wild plant not need preclude its eventual success as

domesticated crop, however. Yields of many wild annual crop pro-
genitors were low originally. Teosinte, the ancestor of corn, yields 305
kg of seed per ha (272 lb/ac) in the wild and 1254 kg/ha (1120 lb/ac) if
cultivated. However, since approximately 50 percent of the harvested
seed is inedible, these yields translate to only 152.5 to 627 kg/ha (130
to 560 lb/ac) of edible portion.[15] Similarly, a field of mixed wild emmer
and barley, ancestral relatives of modern wheat and barley, in eastern
Galilee yielded from 500 to 800 kg/ha (450 to 720 lb/ac).[16] It is quite
possible that perennials could also show dramatic increases in seed
yield with breeding.

Factors Limiting Seed Yield in Perennials. In general, annual crops yield
more seed than do most perennials. Does the fact that a species is
perennial per se set a limit on its seed yield? While it is true that a
perennial must devote some resources to constructing and maintaining its permanent organs, an expense an annual does not bear, it is
not clear that high seed production and perenniality are in any kind
of strict trade-off relationship within a plant. Some scientists have
theorized that just such a trade-off should exist, so that as one bred
to increase seed yield in a perennial plant, energy devoted to the
overwintering structures would decrease. This now appears to be an
oversimplification, and it may be possible to achieve relatively high
seed yields in some perennial plants. Increasing seed production is
probably more complex than merely shifting allocation of resources
and energy from roots and stems to reproductive structures. Several
studies have indicated little or no vegetative cost to future reproduction or differences in survival as a result of increased reproduction in
the present.[17] These results indicate that, for some long-lived perennials at least, an increase in reproduction may not result in the complete loss of the ability to overwinter. This is promising for the effort
to breed for higher seed yield without losing a plant's perennial nature.

CROP MIXTURES

There is a great deal of literature summarizing the ecological and agronomic benefits of crop mixtures. These benefits include overyielding, internal supply of fertility, and adequate management of weeds,
insect pests, and plant diseases. Moreover, there may be compelling
economic or dietary reasons for maintaining mixed cropping systems.

Overyielding. Overyielding in mixtures (Question Two) typically results when crop species differ spatially or temporally in resource use.[18] Compatible crops can differ morphologically, phenologically, or in the types of resources needed. Where crop varieties have been grown together for centuries, compatibility has increased through coevolution. The maize-bean-squash polycultures of traditional Mexican agrarian cultures show signs of such coevolved compatibility.

Agriculturalists frequently employ the Land Equivalent Ratio (LER) as one way of expressing the yield advantage of intercropping over monoculture. The LER, the sum of the fractions of the various intercrops relative to their yields in monoculture, represents the land area required to obtain from monocultures the yield obtained in polyculture.

If the LER is greater than 1, then there is overyielding. A simple illustration may help explain overyielding. Imagine a three-component polyculture, a triculture, consisting of three seed crop species, A, B, and C. In monoculture plantings, the three crops yield 2000, 3000, and 4000 kilograms per hectare, respectively. In the triculture, species A yields 800 kilograms per hectare, species B yields 1200 kilograms per hectare, and species C produces 1500 kilograms per hectare. The LER would then be calculated:

$$LER = 800/2000 + 1200/3000 + 1500/4000$$
$$= 0.40 + 0.40 + 0.38$$
$$= 1.18$$

In other words, to achieve in monoculture the yields obtained in polyculture, a farmer would have to plant 18 percent more land area to monoculture. Thus, intercropping these three crops will be economically advantageous to a farmer if the yield advantage is high enough and exceeds any increased costs of managing the intercrop. If the intercrop and single crops with which it is compared are grown at uniform overall density, then the LER value calculated is the same as the Relative Yield Total (RYT).

If the LER equals 1, then the various yields harvested from the intercrop could have been obtained from the unit area planted to single crops, each occupying an appropriate fraction of the total area. More simply, yields of each of the three crops in polyculture would average 33 percent of their yields in monoculture. When the LER equals 1, the overall yield per unit area of the intercrop is never greater than that of the most productive single crop. In this case, farmers can expect no yield advantage by planting mixed fields instead of monocultures. If the LER is less than 1, then an intercrop is at a disadvantage relative to a monoculture. This might occur if members of one crop species inhibit the growth of the other crops more than they inhibit the growth of members of their own species.

Causes of Overyielding. Conceivably, many factors can lead to overyielding.[19] Some of these factors relate to the spatial configuration of the different crops. First, crops may be released from competition for light if canopies of component crops occupy different vertical layers, with tall crops tolerant of strong light and shorter crops requiring shade and/or relatively high humidity. Examples of this type of cropping system are mixtures of date palm with citrus and vegetables in African desert oases, and shade trees over cocoa or tea in the tropics. Or one crop may produce vertical leaves whereas another, shorter crop produces prostrate leaves. Second, crops may segregate spatially over patchy soils if they have different microhabitat requirements. Third, roots of different species may explore different soil layers, as occurs in mixtures of vegetable and tree crops.

Other factors relate to nutrient or chemical aspects. Fourth on our list, crop species may have complementary nutrient requirements. For example, mixtures of legumes—or other nitrogen-fixers, such as alder (*Alnus* species)—with nonlegumes typically overyield, especially in soils where the nitrogen supply is limited. Another example is mixtures of crops in which species have strongly contrasting nutritional requirements or uptake abilities. Fifth, differences in the length of growing periods or in seasonal periods of nutrient uptake among crops can promote overyielding. Sixth, the presence of different species can change or enhance the activity of rhizosphere flora. A more active rhizosphere may thereby stimulate nutrient uptake and plant growth and lead to an LER much greater than 1.[20] Seventh, if an allelopathic crop is autotoxic—that is, if it harms members of its own species more than it harms other members of an intercrop—it may grow better in a mixture.

Another set of factors relates to physical support or protection among plants. An eighth mechanism that can lead to overyielding involves physical support whereby an erect species provides support for a climbing species and thus more light is available to the vine crop. An example is the familiar maize/pole bean mixture planted in the tropics. Ninth, one crop may provide considerable physical protection to others, as, for example, a tall, robust species that can act as a windbreak for more delicate species. In Australia, citrus may be grown with apricots (*Prunus armeniaca*) to provide protection from wind. Similarly, in Taiwan, young vegetables are often protected from wind by planting them within mature stands of rice.

Tenth, the presence of weeds, pests, and diseases can result in overyielding if an interaction reduces the competitive ability (by reducing shading leaf area, for example), but not the yield, of a crop species, or where the mixture itself lowers the incidence of such harmful organisms.

The most successful multiple cropping systems occur in situations in which crop niches differ somehow. For example, plants may have similar growth forms but mature at different times of the year, as in sorghum/millet intercrops in Nigeria. Alternatively, crops may have different growth forms that allow overyielding. Examples are annuals grown under tree crops, annuals beneath such long-duration crops as sugar cane, or mixtures of tall and short annual crops, such as maize and cassava or maize and peanut.

Intercrops of Perennials. Much of the intercropping involving herbaceous perennial plants is for forage production in pastures. Pasture intercrop studies have shown that the less vigorous component of a perennial intercrop tends to become suppressed with time unless the trend is checked by management. This suppression occurs when the intercrop components differ in competitive ability, because uneven sharing of resources becomes accentuated with time. For example, a perennial grass/legume mixture frequently recommended for the northern United States and Canada is smooth brome (*Bromus inermis*) and alfalfa. By altering the frequency, seasonal timing, and height of hay cuttings, favorable ratios of the two species can be managed for good pasturage.[21]

Several other processes can also occur in intercrops to change the ratios of the component species. One species can be suppressed by pests while another species thrives, or dominance may shift as soil nutrient status changes. For example, legumes may dominate a grass/legume mixture when nitrogen is the factor most limiting growth. With an increase in soil nitrogen, however, grasses may come to dominate such mixtures by growing taller and shading the legume. The relative aggressiveness of a species in a given mixture varies greatly in response to microenvironmental conditions, so that one species may dominate in wet conditions and another in dry. Such shifts may provide the benefit of resilient yields, but they also mean that the farmer must be flexible with management practices.

Intercropped plants can interact in one of two ways. Some interactions between plants are competitive—that is, one species hinders the performance of its neighbor such that overall growth in the field is lowered. Other interactions, such as nitrogen sharing between legumes and grasses, can be positive, leading to improved performance in fields. A clever polyculture farmer can design planting arrangements to maximize the frequency of beneficial contacts and minimize the frequency of suppressive contacts between components in intercropped fields.

Polycultures and Nutrient Balance. The correct mix of species in polyculture can maintain soil fertility (Question Three) by tapping not only the benefits of nonoverlapping resource use, but also the ways species in mixtures benefit one another directly. In particular, legumes benefit other species in mixtures through their special ability to fix atmospheric nitrogen. The nitrogen fixed by their symbiotic bacteria becomes available to other species in the mixture when the legumes' roots decay or via direct mycorrhizal transfer. This biological nitrogen becomes available to the system more gradually than does nitrogen from synthetic fertilizer.

Despite sometimes lower yields than are obtained with synthetic nitrogen, legume nitrogen can prove both energy-efficient and cost-effective. Studies have consistently shown higher dry-matter yields in grass/legume mixtures than in grass monocultures.[22] In "How Much Dinitrogen Fixation Is Required in Grazed Grassland," J. E. Sheehy has calculated that a legume component of 10 percent should be sufficient to maintain grazed pasture. On a nationwide scale, legume nitrogen could conceivably replace synthetic nitrogen fertilizer, albeit with some reduction in yield.[23]

Similarly, mycorrhizae may mediate many beneficial nutrient exchanges among both wild and crop plants (see Chapter 3). Researchers have also found that mycorrhizal associations between plant roots and soil fungi are crucial for phosphate accumulation in prairies and other natural ecosystems.

Polycultures and Weed Management. With regard to the control of weeds (Question Four), we know that some kinds of annual crop polycultures can provide considerable weed control.[24] Perennial mixtures should enhance this effect. They can incorporate overlapping growth periods to enable crops to block sunlight or usurp soil nutrients before weeds take hold. They also provide a continuous groundcover to suppress weed germination. Another biological mechanism of weed control available with perennial mixtures is allelopathy, which would be especially valuable during the vulnerable establishment phase in the first year of a perennial agriculture. An allelopathic crop component, or a low-growing cover crop such as clover, vetch, or alfalfa, could help suppress weeds until the perennial canopy became established.

Of course, not all plants with which a crop shares a field are "weeds." It is possible that some wild cohabitants are beneficial to the crop because of the associated mychorrhizal fungi they support. Others, simply by adding diversity, may reduce levels of harmful insects and diseases on crops (see below). Lastly, where domesticated plants are grown near their wild ancestors, the ancestors can contribute important adaptive genes to their crop relatives.[25]

Polycultures and Insect Pest Control. Research on the effects of polyculture cropping systems on insect pest control (Question Four) confirms that diversity can have many benefits. In general, polycultures tend to reduce densities of insect pests relative to monocultures. A variety of mechanisms can operate within polyculture cropping systems to help control detrimental organisms. First, the presence of a diversity of crop species in a field reduces the ability of pests to locate and move among their host plants. Diversity can create chemical interactions and masking odors that can interfere with colonization, reproduction, and feeding efficiency of phytophagous insects. Second, a polyculture environment often attracts greater than expected numbers of predators and parasitoids. Third, a polyculture can include a trap crop, which attracts pests away from other crop components.

Because of structural diversity, crop mixtures create more ecological niches for harboring beneficial fauna. If richer species diversity favors beneficial carnivores, pest management will gain.[26] Mixtures that include plants that provide pollen and nectar for such beneficial insects as hoverflies (Syrphidae) or parasitoid wasps (Ichneumonidae, for example) gain improved natural pest control. Polycultures can also benefit from plants that sustain populations of neutral insects that serve as alternate food for predators or parasites. These plants help maintain year-round populations of beneficial organisms. Intercrops may modify the crop microclimate (for example, by increasing shade), rendering it unfavorable for insect herbivores.

Documented examples of pest reduction in diverse, intercropped systems abound.[27] Studies involving beans (*Phaseolus* species) discovered that weeds can help protect against insect pests. Bean plots diversified with weeds had fewer adults and nymphs of *Empoasca kraemeri* (Homoptera: Cicadellidae), the main bean pest of the Latin American tropics, than did weed-free bean monocultures. Similarly, less bean damage by *Diabrotica balteata* (Chrysomelidae) occurred in weedy plots (0 percent, 25 percent, and 50 percent of weeds removed) than in relatively weed-free plots (75 percent and 100 percent of weeds removed).

Several other studies involving pests on cabbage family crops found that planted intercrops could reduce pest damage. As reported in their article "Potential of Intracrop Diversity for the Control of Brassica Pests," M. S. O'Donnell and T. H. Coaker grew strips of clover between rows of Brussels sprouts plants so that only 25 percent, 50 percent, and 75 percent of the ground was covered. Using a clover intercrop in this way reduced the number of eggs laid by the cabbage root fly (*Delia radicum*) through the presence of too much nonhost vegetation. In a 1972 article in *Oecologia*, J. O. Tahvanainen and Richard Root demonstrated that monocultures of collards were colonized more rapidly, and experienced greater feeding damage, by the collard flea beetle (*Phyllotreta cruciferae*) than were stands interplanted with tomatoes and tobacco. In a 1975 article in *Philippines Entomology*, R. P. Buranday and R. S. Raros suggested that a 50 percent groundcover was most effective in lowering diamond-back moth eggs on cabbage when tomatoes were used as the intercrop. Intercropping brassicas with French beans and ryegrass (*Lolium perenne*) reduced infestations of the cabbage aphid and cabbage root fly by more than 60 percent compared with those on brassicas in pure stands.[28] Predation was not an important factor in this particular study. Here, the increased movement of flies in mixed plots led to a more rapid departure from mixtures. Therefore, correspondingly fewer eggs were laid.

Intercropping can also reduce pest incidence via disturbance of egg-laying behavior, as odors from an intercropped species can deter or repel certain pests. For example, J. O. Tahvanainen and Richard Root also found that tomato and ragweed (*Ambrosia artemisiifolia*) extracts interfered with host-finding by collard flea beetles.

Interestingly, different types of pests respond differently to polyculture plantings. As reported in his 1980 article in the *Journal of Applied Ecology* and his 1981 article in *Ecology*, Stephen Risch observed that populations of generalist (those that feed on a variety of plant species) chrysomelid beetles increased in polycultures of maize, beans, and squash, whereas specialist (those that feed on just one or a few plant species) beetles tended to decrease in number relative to monocultures. Similarly, in his 1983 dissertation, David Andow reported findings that herbivores that specialize on a single species were more likely to decrease in diverse systems than were generalist herbivores, which were more likely to increase. Presumably, polyculture favored generalists because these species could more readily switch among hosts and were less dependent upon specific chemical cues than were specialists.

Perennialism may also contribute some degree of pest control to a cropping system. Natural enemies of pests may sustain themselves better in perennial than in annual crops because the infrequent cultivation of perennial crops results in less disturbance of the soil and more stability, and the plants provide more structural diversity than do fields of annual crops.

Polycultures and Disease Control. It is well established that disease problems in agriculture oftentimes result from extreme genetic uniformity of crops.[29] Thus, an increase in genetic diversity created by planting mixtures of species, and mixtures of genotypes within species, can reduce the spread of some plant diseases. Therefore, polycultures may hold an advantage over monoculture when it comes to disease control. Although relatively little research has explicitly examined disease patterns in polyculture (Question Four), some control has been documented both within mixtures of disease-susceptible and disease-resistant genotypes[30] and within mixed species fields relative to monocultures.[31]

Quite simply, replacing susceptible plants with resistant ones reduces the amount of tissue vulnerable to infection. This, in turn, reduces the amount of inoculum (spores) available for subsequent dispersal. The increased distance between susceptible plants increases the average distance that inoculum has to travel between susceptible individuals or populations. The mere presence of resistant or non-host plants should also help to reduce disease levels by interfering with and intercepting the movement of inoculum or disease-spreading insects between susceptible individuals.[32]

Jeremy Burdon has summarized in his book *Diseases and Plant Population Biology* the results of more than 100 studies using two-component mixtures representing ten different host-pathogen combinations. Incidence of diseased plants in mixtures was always less than in the most affected monoculture. Moreover, in the great majority of cases, the severity of disease in mixtures was much less than that predicted by the levels of disease occurring in monocultures of the component lines.

Other Aspects of Polycultures. Another important consideration is economic stability as an indirect consequence of the biological stability that can result from polyculture farming. In a practical sense, any mixture permits a farmer to hedge his or her bets by not staking all fortunes on the success or failure of a single commodity. Generally, if one crop does poorly in a given year due to climatic or pest effects, the other crops will compensate and yield better. This compensation should help dampen year-to-year fluctuation and thereby confer overall stability of both yield and income to growers. Alternatively, crops may differ greatly in their monetary value. Hence, it may be economically feasible to accept a lower yield, as a result of intercropping for ecological sustainability, of a particularly high-value crop.

Lastly, a farmer may need to grow crop mixtures for dietary reasons. For example, complete human and livestock diets require both grass and legume crops for amino acid complementariness.

DESCRIPTIONS OF THE LAND INSTITUTE STUDY SPECIES

Researchers at The Land Institute are working to create, in effect, a domesticated prairie. The ongoing research projects at The Land Institute can be grouped into three interrelated areas: (1) studies of the prairie as a model for agriculture, (2) development of perennial seed crops for polyculture, and (3) research in perennial polycultures. Early plant breeding efforts involve developing perennial plants into seed crops to be grown in mixtures that will display many of the sustainable features of native prairie communities.

Development of perennial seed crops is a long-term process that moves in most cases from the currently wild perennial to a successful crop that retains the essential functional attributes of prairie plants. Work at The Land Institute to domesticate perennial seed crops began in 1978 with an inventory of nearly 300 herbaceous perennial species for their suitability to the environment of central Kansas and promise of high seed yield. A second inventory studied the agronomic potential in 4300 accessions (collections) of perennial grass species within the cool-season genera *Bromus, Festuca, Sporobolus, Lolium, Agropyron, Elymus,* and *Leymus.* From these inventories, a handful of perennial species was chosen for potential crop development.[33] Presently, the species chosen represent three approaches in the development of perennial seed crops: domesticating wild native perennials (eastern gamagrass, Illinois bundleflower, and Maximilian sunflower), domesticating a wild introduced perennial (leymus), and creating a perennial version of a widely planted grain crop (grain sorghum) via a wide cross with a perennial relative.

Eastern gamagrass (*Tripsacum dactyloides*) is a warm-season grass native to the region stretching from the southeastern United States and Great Plains southward to Bolivia and Paraguay. It can cross with corn and may have been important in the evolution of corn. Although eastern gamagrass is acclaimed as a select forage, it shows much promise also as a human grain crop. Gamagrass grain is both tasty and nutritious, being about 27 percent to 30 percent protein and 7 percent fat.[34] The ground seed has baking properties similar to those of cornmeal. This species begins flowering in central Kansas in late May and seed harvest begins in July. The major limitations of eastern gamagrass as a grain crop are low seed yield and enclosure of the grain in a hard fruit case. As a possibility for improving seed yield, The Land Institute has incorporated into its breeding program a gynomonoecious (or pistillate) variant, forma *prolificum,* which produces exclusively female flowers and thus has the potential to increase seed yield considerably.[35]

Leymus (*Leymus racemosus*), a member of a group of species collectively called wildrye, is within the tribe Triticeae, which also contains wheat, rye, and barley. It is a rhizomatous, cool-season grass native to sandy soils in Bulgaria, Romania, Turkey, and parts of the Soviet Union. It can form weak perennial hybrids with wheat, rye, and barley. Grain of this and related species was eaten by Asian and European people historically, especially in drought years when annual grain crops faltered. As is typical of cool-season plants, leymus displays most of its growth in late autumn and early spring. Flowering in plots at The Land Institute takes place in May and seeds mature by late June.

Grain sorghum (*Sorghum bicolor*), a native of the African continent, is a successful seed crop in the southern Great Plains. It is weakly perennial in tropical regions but is killed by frost at higher latitudes. Johnsongrass (*Sorghum halepense*), a weedy relative of cultivated sorghum, is, in North America, a noxious weed that overwinters by production of rhizomes. The Land Institute is exploring the feasibility of converting grain sorghum from an annual to a perennial growth habit by transferring the ability to produce winter-hardy rhizomes from Johnsongrass. With care taken to avoid creating a new weed, the goal is to select a good grain type that also can produce enough rhizomes to be winter-hardy. The ease of making this transfer will depend on the number of genes controlling the production of rhizomes and whether overwintering ability is genetically associated with poor agronomic characteristics.

Illinois bundleflower (*Desmanthus illinoensis*) is a nitrogen-fixing legume that forms a deep taproot in its first year. It is native to the Great Plains, and its range extends north into Minnesota, east into Florida, and as far west as New Mexico. It grows best during warm weather, flowering from late June onward; the small seeds are borne within clusters of brown legumes beginning in late July. The nutritional quality of the seeds (38 percent protein and 34 percent carbohydrate),[36] and the edibility of seed of other *Desmanthus* species in Mexico,[37] suggest great potential as a human seed crop. This species appears capable of fixing appreciable amounts of atmospheric nitrogen. As with other wild legumes, the pods open (dehisce) and disperse their seeds upon maturity. A nonshattering type identified in 1988 has been crossed with accessions displaying consistent high seed yield, and these crosses were planted for further evaluation in 1990.

Wild, or Maryland, senna (*Cassia marilandica*) is a legume native to the southeastern region of the Great Plains. Flowering in Kansas takes place from late August to early September and produces racemes of insect-pollinated yellow flowers that become brown-black legumes later in the fall. Plants may grow up to two meters tall. *Cassia marilandica* produces thick, deep roots, but it probably does not form symbioses with nitrogen-fixing *Rhizobium* bacteria.[38] Edibility of its seeds for humans or livestock is unknown. Although The Land Institute is not working to develop this species as a seed crop, much information on its long-term patterns of seed yield has been gathered to address the biological question of whether an herbaceous perennial can produce a sustained, high-seed yield.

Maximilian sunflower (*Helianthus maximilianii*) is a perennial member of the Compositae, or sunflower, family native throughout the Great Plains. Its range extends eastward to Maine and North Carolina and westward to Texas and the Rocky Mountains. Flowering begins in late August and September; seeds begin to ripen in October. In addition to its potential value as a seed or oil crop (seed is 21 percent oil)[39] Maximilian sunflower appears to inhibit weed growth allelopathically and may therefore be especially important during the establishment phase of a perennial polyculture.

THE LAND INSTITUTE'S RESEARCH RESULTS

The Land Institute's research is directed toward developing agriculture that mimics the structure and sustainable function of the prairie. The four basic questions outlined earlier form the framework for this research, but in addition, studies of the prairie itself also direct the perennial polyculture work.

PRAIRIE STUDIES RELEVANT TO SUSTAINABLE AGRICULTURE

The purpose of studying the prairie is to learn more about what makes it self-sustaining in the face of a drastically fluctuating climate and what enables it to maintain its own fertility, even build fertile soil, despite a sometimes harsh climate. Even though researchers expect to incorporate the processes that provide sustainability simply by borrowing the prairie's structure, it is still necessary to understand as much as possible about the prairie itself. Knowledge of how the prairie responds to drought and other stresses helps in incorporating the essential features, such as the proper mix of species with early-versus late-season flowering or species with deep versus shallow roots. It helps researchers capture the functioning of the prairie by fine-tuning the structure to better mimic the prairie. Similarly, the more that is known about how the prairie maintains its own fertility—in particular, about how the species mix affects this process—

the better the chance of creating a model that functions the same way. Since researchers want a mimic at the structural level, they are most interested in such questions as what proportion of the vegetation is nitrogen-fixing legumes and how that proportion varies with soil types and other physical factors. This sort of knowledge will allow farmers to tailor the species mix to suit different fields.

The prairie ecosystem provides the model The Land Institute is striving to imitate with perennial polyculture. Maximum plant productivity on prairie sites may represent the sustainable bench mark for a sunshine agriculture. The studies of native prairie revolve around such questions as: What is the vegetation structure of a typical prairie community? How does the vegetation structure change across soil types and in wet versus dry growing seasons? Since stable crop diversity is what researchers hope to accomplish, it is necessary to untangle the factors that allow, maintain, or even promote plant species diversity in natural communities. How does diversity, within or between species, contribute to the prairie's resilience? Specifically, where and under what conditions does the prairie feature warm-season grasses, cool-season grasses, legumes, and composites?

Studies thus far across prairie sites ranging from poor to richer soils have shown some consistent patterns.[40] Under extended drought, overall biomass production declines, and most species are diminished but do not disappear from the plant community. Some species decline to a greater extent than others. Different types of plants display different mechanisms for surviving drought. Many prairie grasses produce deep, fine roots for acquiring subsoil moisture. Many perennial forbs, on the other hand, rely on deep taproots. Certain spring-flowering forbs and cool-season grasses avoid summer drought by completing their reproduction in June.

As stated above, the purpose of The Land Institute's research is to create a polyculture of perennial seed crops that functions like a prairie ecosystem. But the vegetation structure of the prairie, the model to be mimicked, varies across soil types. For example, four sites studied at The Land Institute represent a productivity gradient determined by combinations of moisture and nutrient levels (see Tables 5.1 and 5.3). Hence, the sites differ in the mix of species that constitute the respective plant communities. Since the perennial crop candidates are within the categories of C_4 grasses, C_3 grasses, legumes, and composites, it is most important to examine how these groups are represented on a range of prairie sites (see Table 5.2). The vegeta-

TABLE 5.1

Descriptions of Four Prairie Research Sites

Site	Bedrock	Percent slope	Aspect	Description of soil
1	Sandstone	5–7	SW	Deep, high nitrogen, high organic matter
2	Sandstone	5–7	W	Deep, high nitrogen, high organic matter
3	Shale	27–43	W	Shallow, low nitrogen, low organic matter
4	Sandstone	13–16	S	Shallow, high nitrogen, low organic matter

TABLE 5.2

Percentage Composition of Four Prairie Research Sites

Site	C_3 graminoids[a]	C_4 grasses	Legumes	Composites
1	19.6	75.8	0.0	4.7
2	12.5	79.3	0.0	8.1
3	0.0	60.0	16.6	8.2
4	2.3	76.6	1.6	17.8

[a]Graminoids = grasses and sedges.

tion of sites 1 and 2 is composed of more than 90 percent grasses, primarily warm-season, although cool-season grasses can be as high as 20 percent. In contrast, site 3 is only 60 percent grasses, almost exclusively warm-season, and composites average only 8 percent of biomass; but legumes average nearly 17 percent. The relative amount of legume biomass suggests the percentage needed to support a sustainable agriculture composed of grass/legume mixtures. Lastly, site 4 is about 80 percent grasses, but with 18 percent composites and very few legumes.

The way these prairie plant communities are structured taxonomically across soil types is interesting. C_3 and C_4 grasses dominate deep, fertile soils whereas a dry, infertile site supports a high proportion of legumes but virtually no C_3 grasses. A fertile but dry site features few legumes and C_3 grasses, but many composites. Note, too, that the four plant groups—C_3 and C_4 grasses, legumes, and composites—account for all the vascular plant biomass at sites 1, 2, and 4, whereas other plant families (for example, Liliaceae, Labitae, and Umbelliferae) together account for 16 percent of the biomass, on average, at site 3.

Species richness, diversity, and evenness, then, are generally higher on sites of poorer soil quality (see Table 5.3). Different assemblages of species, however, contribute to this diversity depending on the factors that cause low productivity. The low productivity of site 3 is due to low soil moisture and nitrogen fertility, and thus this soil favors a legume component that may be as high as 26 percent during some periods. Site 4, however, has dry soil but is fairly high in nitrogen. Here, legumes are not favored, but instead composites represent a fairly high proportion of the vegetative biomass of the site.

The results suggest that a number of related mechanisms control species diversity on prairies. Different species appear to be adapted to different nutrient levels or microhabitats, as they tend to segregate in different sites within the prairie. Above-ground, taller species tend to eliminate shorter species on rich soils. Below ground, roots compete for pools of moisture and nutrients. The general observation that poor soils tend to display greater species diversity indicates that poor soils will allow, or may promote, greater species diversity.

This variety of community makeups on native prairie sites sug-

TABLE 5.3

Ranges of Some Community Attributes Measured at Four Prairie Sites, 1986–90 (Values for Site 4 Are for 1990 Only)

Site	Peak biomass[a] (g/m²)	Richness[b]	Evenness[c]	Diversity[d]
1	387–1078	7.1–9.6	0.43–0.49	2.68–3.23
2	327–535	8.2–11.8	0.47–0.58	3.15–4.21
3	190–371	6.9–10.2	0.63–0.68	3.50–5.02
4	293	11.8	0.64	4.76

[a]Figures expressed in grams per square meter can be multiplied by 9 (g/m² × 9) to give approximate pounds-per-acre equivalent.
[b]Average number of species per quarter-meter-square sample frame.
[c]A measure of equitability among species. Maximum evenness = 1.
[d]The exponential of the Shannon Index (H′), which factors in the richness and evenness components. This value represents the number of equally common species required to produce the value of H′ given by the sample.

gests that researchers should experiment with a range of designs for polyculture. In some cases, a polyculture might emphasize warm-season grasses, or legumes, or cool-season grasses, depending on soil conditions.

Of course, the prairie has a very different purpose for existence from a grain field. Prairie plants have evolved to survive drought, fire, and large grazers. Consequently, they emphasize growth below ground and recruitment from seed is uncommon. It is therefore not surprising that Native Americans of the Plains favored the roots of many prairie plants as staples. Seed production of most prairie plants represents a very small percentage of a plant's energy outlay. This presents an open question. As perennials are domesticated as seed crops, will they, with improved seed yields, still perform their beneficial ecological roles in polyculture?

QUESTION ONE: SEED YIELDS OF PERENNIALS

The work with perennial polycultures began with an inventory of herbaceous perennials from which five species were chosen for further, long-term observations of plant growth and seed yield. Conceivably, over time, seed yield may increase with plant size, decrease as plants age or deplete some critical soil nutrient, or oscillate predictably or irregularly. In some species, flowering, and therefore seed production, increases with plant size to a maximum constant output. Other species peak early, then decline in seed yield thereafter. No single model appears adequate to predict the long-term patterns observed in perennial monocultures maintained without benefit of fertilizer or nitrogen.[41] This is important because even if yield levels are not constant, they at least must be predictable.

Yields in Land Institute Plots. Some of the highest yields observed in plots at The Land Institute have been in Illinois bundleflower, with a plot that produced relatively high yields during the first three years, including a peak of 197 grams per square meter (1970 kilograms per hectare, or 1760 pounds per acre), and in a wild senna plot, in which yields oscillated year to year but averaged 209 grams per square meter (equivalent to 1870 pounds per acre), in the second year. Peak yields of leymus, eastern gamagrass, and Maximilian sunflower have been somewhat lower (see Table 5.4). In general, yield tends to decline in subsequent years from an initial peak. One notable exception

TABLE 5.4

Highest Seed Yields in Selected Experiments Involving Perennial Seed Crop Candidates at The Land Institute

Species	Type of experiment	Duration (years)	Highest yield[a] (g/m²)	Year of highest yield[b]
Illinois bundleflower	Density	5	197	1st
	Germplasm	2	134	1st
	Polyculture	3	163	1st
Eastern gamagrass	Germplasm	3	24[c]	3rd
	Polyculture	3	20[c]	2nd
	Effects on soil	2	25[c]	2nd
Wild senna	Germplasm	5	121	4th
	Density	5	209	2nd
Leymus	Yield	5	51	2nd
	Polyculture	3	34	2nd
	Germplasm	2	83	2nd
Maximilian sunflower	Density	5	30	1st
	Comparison with annual	2	77	1st

Source: Jon K. Piper and Douglas Towne, "Multiple Year Patterns of Seed Yield in Five Herbaceous Perennials," *The Land Institute Research Report* 5 (1988): 14–18.
[a]Figures expressed in grams per square meter can be multiplied by 9 (g/m² × 9) to give approximate pounds-per-acre equivalent.
[b]Leymus and gamagrass are vegetative in the first year if established from seed.
[c]Estimate based on seed yield as 27 percent of uncleaned fruit-case yield.

occurred in a planting of eastern gamagrass in which the yield of pistillate plants increased fourfold from the second year in the field (the first year they produced seed) to the third year.

A next step is to compare these base-line monoculture yield data with yields achieved for these species in polyculture. This comparison addresses the issue of whether complementariness among compatible species can compensate for declining yields under low-input conditions.

Selection for Higher Yields in Perennials. Selection and domestication of genotypes for use in perennial polyculture are inherently more complex than selection of varieties for monocultures. The process is complex both because it involves initial domestication of wild plants and because it requires adaptation to mixed-species plots. Initially, scientific plant domestication involves germ plasm collection and evaluation, then selection for stable high seed yield, harvestability, and suitability for use as a human or, perhaps, as a livestock food. Harvestability refers to the practical aspects of getting seed from the plants in the field to the point where it can be processed for food. Characteristics that make seed more harvestable include seeds that don't easily shatter (fall off the plants), but are easily threshed, and flowering stalks that resist lodging (falling over). Thus several characteristics must be selected for simultaneously. Furthermore, these characteristics need to be evaluated for several years in the life span of a perennial species. First-year stalks could be sturdier or more fragile than third-year stalks, for instance, a trait that would affect the lodging potential.

The process becomes more complex because a plant with a particular genotype may respond differently to different species or varieties as neighbors. The plant breeder must therefore evaluate each genotype in mixtures of species and in different arrangements of species mixes. The interactions between genotypes and species and cropping arrangements may be unpredictable from responses in monoculture. Varieties that succeed in polyculture must complement one another in nutrient and water use and have compatible canopy structures and management requirements.

Screening for Desirable Traits. The breeding program at The Land Institute is in the preliminary stages of domesticating perennial seed crops. Peter Kulakow has been assembling germplasm from natural plant populations and then screening these populations to describe genetic variation patterns and to identify a range of genotypes that warrant further testing. Germ plasm collections representing the species' geographical distributions are in various stages of completeness for Illinois bundleflower, eastern gamagrass, and leymus. By the end of 1990, 16 accessions of leymus and its relatives, more than 250 accessions of eastern gamagrass, and about 150 accessions of Illinois bundleflower had been collected, documented, and/or evaluated in the ground at The Land Institute. The next step, describing the available variation in plant growth and agronomic performance within each species, is also under way for these three species. Favorable genotypes are then tested in a set of cropping systems that includes monocultures and polycultures. This work is venturing into fairly new territory in plant breeding. So far, the appropriate set of procedures for use in breeding for low-input perennial polycultures has not been defined. This work will therefore have a double benefit: to help develop those procedures, as well as to develop new crops.

Despite the fact that the screening program is still in its early stages, some collections already show promising traits. For example, some collections of eastern gamagrass show consistent (year to year and across two locations) resistance to or tolerance of various leaf diseases. Some accessions of both gamagrass and Illinois bundleflower display relative drought hardiness or high seed yield. Non-shattering Illinois bundleflower has been crossed with some high-yielding lines of this species.

Working Toward a Perennial Sorghum. In the study involving a wide cross, interspecific hybrids were made using tetraploid lines of grain sorghum and ten collections of Johnsongrass from Kansas and California. In 1986, sorghum and Johnsongrass were crossed to produce an F_1 generation, which was then crossed with itself to produce an F_2. The F_1 and F_2 generations were grown out in 1987 and evaluated for twenty-two characteristics, including rhizome production and agronomic characteristics, that distinguish cultivated from weedy sorghums. Rhizomes are the key to overwintering ability.

Although a large amount of variation resulted from these crosses, the frequency of agronomically desirable plants was low. Yet variation patterns indicated to Peter Kulakow that desirable perennial sorghum populations could be developed through recurrent selection. Eighty-four percent of the F_2 individuals produced rhizomes. Segregation ratios for the presence and absence of rhizomes supported the idea that one or a few genes determine rhizome production. A series of backcrosses to the cultivated parent, producing hybrid populations with 0 (BC_0), 1 (BC_1), and 2 (BC_2) levels of backcrossing containing 50 percent, 75 percent, and 87.5 percent grain sorghum genes, respectively, were then made to improve the chances of recovering desirable plants. These populations with different percentages of grain sorghum genes were used to assess the level of backcrossing most useful for developing a perennial sorghum. A few plants with 87.5 percent cultivated genes (the second backcross) displayed good agronomic traits, and many plants of this generation produced rhizomes in the field.[42]

The work in 1990 was an evaluation of backcrossed F_3 families for shatter resistance, seed size, yield, and winter hardiness. Among these plants, 88 percent of BC_1 and 57 percent of BC_2 plants had some rhizome production. Seed yields of both backcrosses were more than 150 grams per square meter (about 1340 pounds per acre). Surprisingly, there was a positive correlation between seed yield and rhizome mass in BC_1 plants, indicating that particularly vigorous plants had both high seed yield and rhizome production.[43] Whereas retaining the ability to form rhizomes in backcrossed plants has proved relatively simple, overwintering ability in this plant material has yet to be achieved.

Evaluating Plant Material in Mixtures. As noted above, plant breeding within a polyculture context must take into account the fact that some lines may behave differently when grown in different planting arrangements or with different species for neighbors. These sorts of interactions need to be evaluated before a breeding program for polycultures is designed. If it turns out that no such interactions are found—in other words, the best performing lines in monoculture are also the best in polyculture—it will be possible to simplify the breeding program. This result would mean that the plant breeder could evaluate and select different lines in monoculture and still be able to predict their relative behavior in polyculture. If some lines respond differently in monoculture versus polyculture, however, evaluation and selection must take place in polyculture. Moreover, since The Land Institute is working with perennials, interactions may differ in different years, and so the breeding program must also account for year-to-year variation.

In 1988, assessments of Illinois bundleflower and eastern gamagrass germ plasm variation in different cropping systems began with the growing of collections in monoculture and biculture plantings. In this experiment, twenty-eight collections each of Illinois bundleflower and eastern gamagrass were established in monocultures and in a biculture with three replications for each cropping system. Data were collected on growth form, presence of disease, flowering dates, and yield-related characteristics.

In 1989, the second year of the study, several characteristics of eastern gamagrass, basal area (an index of plant size), plant height, number of reproductive tillers, yield, and flowering time differed between cropping systems. On average, gamagrass accessions flowered earlier in biculture plots. For the other variables, accessions performed better in monoculture than in biculture.

In Illinois bundleflower, plant height, height of the lowest bundle of seed pods, spring vigor, and days to first flowering all varied according to cropping system. Illinois bundleflower plants were taller and flowered later on the average in biculture.

In 1990, eastern gamagrass had few traits influenced by cropping system. Whereas in 1989 plants were smaller and produced less seed in biculture, in 1990 plants in biculture had smaller basal areas only. Generally, Illinois bundleflower accessions were much more sensitive to cropping system. In 1989, for example, plants in biculture were taller, flowered later, and were less vigorous in the spring. In 1990, plants yielded less in biculture than in monoculture.

Within each species, accessions differed from one another in many respects. For example, in 1990, the experiment's third year, gamagrass accessions differed in basal area, number of reproductive tillers, seed yield, and disease severity. And, whereas in 1989 bundleflower accessions differed in all variables except days to first flowering, in 1990 accessions differed from one another in vigor and height only.[44]

Relative performance by different accessions may depend upon the particular cropping system in which they are evaluated, a phenomenon called accession by cropping system interaction. Thus, it is important to look at any accession by cropping system interactions because accessions that rank highest for certain characteristics in monoculture may not rank highest in certain mixed cropping systems. When such interactions occur, the best performing accessions may differ among cropping systems, and selections should therefore be made within the intended cropping system.

For both species, then, plants from different source populations behaved differently in different cropping systems. From this result it can be concluded that breeding for certain traits in these crop candidates should be pursued in polyculture plantings. The subject of overyielding in this monoculture/biculture experiment is discussed in the next section.

Another study showed that Illinois bundleflower, eastern gamagrass, and leymus appear to be compatible for several years in polyculture, although neighbor species can have dramatic effects on phenology, growth, and seed yield of individual plants. The following is an account of a three-year (1987–89) study of these species in a mixed planting.[45]

Harvest phenology of leymus did not depend on neighbor differences during the first two years. In the third year, however, leymus seed from plants next to Illinois bundleflower ripened faster than leymus seed from plants neighbored by other leymus. For example, by the end of June, 77.5 percent of seed from leymus neighbored by Illinois bundleflower individuals was harvested whereas only 53.0 percent of seed from leymus plants growing next to other leymus was harvested by that date. There were no differences in leymus plant height or number of reproductive tillers among neighbor treatments in any year. In the second year, however, leymus yielded better when neighbored by its own species than when next to gamagrass, but leymus yield was not affected by having Illinois bundleflower as a neighbor. Neighbor treatment did not influence leymus yields in the final year.

Both harvest phenology and height of eastern gamagrass were fairly insensitive to neighbor treatment during the study. In the second year, however, eastern gamagrass next to either leymus or Illinois bundleflower produced more reproductive tillers than did gamagrass plants next to members of their own species. In the second and third years, gamagrass plants yielded best when neighbored by Illinois bundleflower.

In contrast, Illinois bundleflower was very sensitive to neighbor species, and the effects differed in different years. For example, in the first year, bundleflower plants adjacent to gamagrass plants ripened fruits sooner than did those adjacent to leymus or other bundleflower plants. By the tenth of August that year, 83.4 percent of bundleflower seed had been harvested from plants next to gamagrass plants versus only 55.0 percent of seed from plants next to other bundleflowers. In the last two years, however, seed of Illinois bundleflower plants in the midst of other bundleflowers ripened faster than bundleflower seed in plants next to either grass. First-year bundleflower plants grew taller and yielded better if next to a grass. In the second and third years, however, the growth and yield of bundleflower were not affected by neighbor treatment due to the overwhelming negative effects of drought and insect herbivores on this species in those years.

One of the significant outcomes of this study is that it revealed both positive and negative associations between species and changes in the directions (positive or negative) of interactions in different years. As mentioned earlier, future successful mixtures of perennial crops will have to accentuate the net positive while minimizing any net negative associations.

QUESTION TWO: OVERYIELDING IN PERENNIAL POLYCULTURE

Yet another study, a biculture, attempted to combine two crops with complementary resource use to obtain overyielding. As noted earlier, overyielding is most profound where crops in polyculture do not compete directly for limiting resources. In this case, a series of mono- and bicultures were planted involving wild senna, a perennial legume that does not fix nitrogen, and Illinois bundleflower, a perennial legume that fixes appreciable amounts of nitrogen. As expected, significant overyielding occurred by the second year. Moreover, this overyielding effect apparently increased over time. In the alternate row component—that is, species were repeated every other row—of the biculture, overyielding rose from 11 percent to 51 percent over three years. The increase was more dramatic in the alternate plant biculture—that is, plant species were mixed within each row—in which overyielding averaged 161 percent by the third year. Essentially, absolute yields declined in monocultures whereas yields in bicultures remained fairly constant or declined more slowly over time.[46]

This biculture study involving the two legume species indicated a clear benefit to a perennial via association with a nitrogen-fixing species. It also showed that polycultures can counteract the trend toward decreasing yields in subsequent years that has been observed in some perennials.

An additional goal of the study in which twenty-eight different accessions each of Illinois bundleflower and eastern gamagrass were grown in a series of monocultures and bicultures was to estimate overyielding. In the second year, the first year in which gamagrass produced seed, overyielding (LER) was calculated in two ways. First, if the LER was based on the *average* yields across genotypes in monoculture and biculture, then the LER equaled 1.25, or a 25 percent yield advantage in biculture. Second, if the LER was based on the *best* yields in monoculture and biculture, then the LER was 1.19, a value similar to that derived from the average yields. In the third year, overyielding based on best yields was 1.08. These positive results demonstrate that overyielding, typical in many polycultures of annual crops, can also occur in perennial polycultures and can hold from year to year.[47]

QUESTION THREE: INTERNAL SUPPLY OF FERTILITY

A likely mechanism for the overyielding observed in the senna/bundleflower biculture described above was the ability of Illinois bundleflower to provide much of its own nitrogen requirements via symbiosis with nodulating bacteria. The mean acetylene reduction rate, an estimate of nitrogen fixation under laboratory conditions, measured in 70-day-old Illinois bundleflower plants was 141 nanomoles per minute.[48] This value is similar to or somewhat higher than values reported elsewhere for this species and also for 68-day-old soybeans.[49] Thus, senna plants in the bicultures probably had more soil nitrogen available for their growth than did senna plants in monoculture, which translated into relatively greater seed production. The results of this experiment indicate that, for the first three years at least, nitrogen fertility was available from the bundleflower to maintain the biculture cropping system.

Another experiment has documented how leymus, eastern gamagrass, and Illinois bundleflower alter the moisture and nutrient regimes of the soils they occupy during the first two years of growth.[50] These plants were grown in replicated monoculture plots arrayed randomly on a common soil type. Hence, any changes seen in soil properties over time would have been due to the effects of the particular plant species growing there. The study correlated some changes in soil variables with periods of growth, flowering, fruiting, and dormancy.

In general, soil moisture declined rapidly within Illinois bundleflower plots, but more gradually within plots of the grass species, and generally remained most favorable in leymus plots. Such a moisture reserve could be available to warm-season components in a polyculture during summer. Inorganic nitrogen concentrations at shallow (10-centimeter, or 4-inch) and intermediate (40-centimeter, or 16-inch) depths declined by the end of the second year in eastern gamagrass plots, but remained higher in plots of leymus over the two years. In early 1991, after three years of growth, gamagrass plots were generally highest in soil organic matter and total nitrogen. Soils of leymus plots were consistently highest in NO_3 concentration. Phosphorus at a 10-centimeter depth generally increased across the study period, a pattern expected if mycorrhizae were active in these plots. Further monitoring of these types of plots is necessary to examine long-term soil nutrient depletion or enrichment by these plants.

Results from this study on plant-soil interactions will help test hypotheses concerning underground activity in perennial seed crops, maintenance of soil fertility, and cropping system effects on fertility and moisture. In addition, the results will suggest optimal planting densities, ratios, and prospects for long-term fertility management of these species within a large-scale polyculture.

Leymus, Illinois bundleflower, and eastern gamagrass have been included in a two-year, LISA-funded experiment at The Land Institute in collaboration with agronomists from Kansas State University and farmers via the Kansas Rural Center in Whiting. The goal of this experiment is to compare the effectiveness of different legumes in contributing nitrogen to grain crops. This is an important question because amounts of nitrogen fixed by legumes and rates of uptake differ considerably among species. Several legume species (alfalfa, hairy vetch, soybean, Illinois bundleflower, and sweet clover) are being evaluated with several different grains (winter wheat, grain sorghum, eastern gamagrass, and leymus). The study is evaluating and quantifying both how much nitrogen the different species supply and the timing of the legume nitrogen availability for grain uptake. At another location, this experiment will be implemented under three tillage systems.

Ideally in an intercrop, the period of maximum nitrogen mineralization would overlap or coincide with the period of maximum uptake in a subsequently planted crop or companion crop. Studying several species should show how much nitrogen is fixed by different legumes and when it becomes available to various grass crops. The study will be particularly useful to The Land Institute because valuable information on perennial crop candidates will be collected in a context that allows direct comparison with crops about which much is already known.

QUESTION FOUR: MANAGEMENT OF WEEDS, INSECTS, AND PLANT PATHOGENS

Weeds. Effective weed control has occurred in two separate experiments at The Land Institute. In one study, a plot of Maximilian sunflower containing controls and rows planted at five densities, weed biomass was significantly reduced in the sunflower plots relative to the control. By the second year, a sunflower density of 3.6 plants per square meter (3 plants per square yard) reduced weed biomass between rows to levels 25 percent to 50 percent of the control levels. In the third year, weed biomass in sunflower rows was 44 percent of weed biomass in the control during May (Figure 5.1). This effective weed control was maintained across years despite changes in the weed community from predominantly annuals in the first year to perennials in the third year.

FIGURE 5.1 Weed biomass within various densities of Maximilian sunflower in May and July 1986.

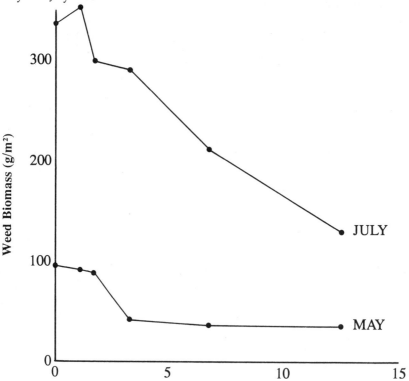

Helianthus Maximilianii--Number of Plants per Meter of Row

SOURCE: Drawn from data in Mark Gernes, "Effect of *Helianthus maximilianii* Density on Biomass of Naturally Occurring Weeds." *The Land Institute Research Report Supplement* 3 (1986): 39–42.

In the second experiment that examined weed growth, a triculture comprising leymus, eastern gamagrass, and Illinois bundleflower at equal densities,[51] species combinations differed in their ability to control weeds. Weed biomass was consistently lower between leymus/gamagrass rows than between either leymus/bundleflower or gamagrass/bundleflower rows across the growing season, despite seasonal changes in species composition of the weed community. In the third year, less weed biomass was produced between rows of leymus and gamagrass than between rows of bundleflower and gamagrass during the summer and fall. In the fall, areas between bundleflower and gamagrass rows also supported fewer weeds than areas between rows of leymus and bundleflower. These results point to eastern gamagrass as the primary weed controller among the three species, probably via shading, although unmeasured underground interactions were also likely important throughout the three-year study. The best mix of perennials for weed management must be worked out in trial polycultures with various ratios of their component species.

Insects. Each of the perennial crop candidates harbors one or more obvious insect or fungal pests, whose effects on growth and yield will have to be measured. For example, Illinois bundleflower flowers and leaves are eaten by a chrysomelid beetle, *Anomoea laticlavia*, which can reduce seed production dramatically. This insect was particularly abundant in 1988. Its high rate of flower predation that year tended to mask any treatment effects on yield in some experiments. Eastern gamagrass appears subject to predation by such generalist herbivores as grasshoppers. By observing how regularly these detrimental organisms appear in various experimental plots, researchers will eventually be able to establish the relative merits of polycultures in reducing the incidence of insect pests relative to monocultures.

Plant Pathogens. Although the two grasses appear less prone to devastation by insects than the legume bundleflower, they are subject to several diseases. Leymus is susceptible to infection by ergot (*Claviceps purpurea*) in wet years, and eastern gamagrass can suffer from several foliar diseases, including leaf rust and maize dwarf mosaic virus.[52] The worse the diseases become, the lower the seed yield. Fortunately, some gamagrass accessions show consistent genetic resistance to diseases in different locations.[53] Additional work to examine plant disease levels within different types of perennial polycultures will be undertaken as successful polycultures are developed.

FUTURE RESEARCH IN PERENNIAL POLYCULTURE

Over the next few years, research at The Land Institute will continue to address multiyear seed yields in perennial crops, developing lines for optimal performance in polyculture, overyielding in polycultures of perennial crops, soil nutrient changes over time and capacity of the system to compensate for losses, and nonchemical management of detrimental organisms. The work thus far has provided data crucial to the design of new perennial polycultures planted in 1991. These perennial polycultures represent the logical next step in the move toward a synthesis of the prairie studies with agronomic studies. Some of these polycultures are mixtures that reflect the proportions of different types of species found on prairie sites. Thus, the research will come full circle as it blends the ecological prairie studies with agronomic polyculture research. Incidence and identity of weeds, insects, and diseases will continue to be monitored in monocultures, and in various mixtures of species, and in plots on different soil types.

The ultimate value of the research at The Land Institute is in its move toward an agriculture based on sunlight, with closed nutrient loops, that uses nature as its model. A diverse and stable agriculture based on mixtures of perennial seed crops would provide numerous environmental and social benefits. In addition to the savings in cost and soil, science itself would benefit as ecologists merge their expertise with that of agronomists to develop new insights and syntheses for agroecology. This would pave the way for agriculture to be regarded more as a biological than as an industrial science. For instance, successful farmers could tailor diverse cropping systems to accommodate heterogeneous landscapes rather than maximize production by streamlining and homogenizing landscapes through the application of synthetic inputs.

THE WORK REMAINING

Although this chapter focuses on the scientific feasibility of prairie-based agriculture in the Great Plains, the ideas can be extended to encompass the feasibility of nature-based agriculture in other regions around the world. Promising results have been demonstrated in four fundamental areas: potential for high-yielding perennials, potential for overyielding in perennial mixtures, potential for maintenance of soil fertility without synthetic inputs, and potential to manage weeds, pests, and diseases biologically. Though the research provides hope for future success, more work remains. A great many of the challenges will be ecological in nature, featuring studies of nutrient cycling, the roles of soil organisms, ecological requirements and roles of different species, ecological succession, long-term stability of yield, and biological management of insects, diseases, and weeds. Some will focus on further understanding of the prairie as model, and some will focus on the agricultural goal of seed production in the context of the prairie model. Another large part of future work involves the challenge of developing successful perennial grain crops.

This emerging field of sustainable agriculture requires the expertise of scientists across the range of ecological and agricultural subdisciplines.

Evolutionary ecologists can contribute to the development of crops appropriate to local climates and soils, sufficiently and consistently productive perennial grains, factors conferring host resistance to pests, and coevolution between crops and other species.

Plant population biologists can contribute insight and methodology to examine factors affecting long-term sustainable yield, characteristics of size- and age-structured stands, and relationships between intraspecific genetic diversity and patterns of resilience and resistance.

Community ecologists' knowledge of interspecific interactions can help in the design of polycultures that maximize beneficial interactions among crop species. More specifically, to promote mutualistic interactions, knowledge of pollination (for example, roles of specialist and generalist pollinators in mixtures), symbiotic nitrogen fixation (for example, how soil environment and vegetation affect soil microbes), and mycorrhizal associations (for example, how species diversity affects mycorrhizal density and the exchange of nutrients between crop species) will be a great aid. Community ecologists can also help find ways to promote conditions that favor beneficial insect predators and parasitoids, manage weeds and diseases, and optimize the competition/facilitation balance between interplanted crops.

Perhaps most importantly, ecosystem ecologists' knowledge of natural ecosystems can help to incorporate into agriculture techniques that favor efficient nutrient capture and cycling, energy flow, diversity, stability, productivity, and resilience.

Plant breeders (using traditional techniques, primarily) will be involved in studying the interplay between population genetic diversity and stability and resistance. Breeding crops for success in polyculture is a largely unexplored field. Part of this effort will involve examining the population structures of perennial plants in nature and mimicking them in agriculture. Breeders might ask: How much diversity exists in nature and what are the factors that maintain it? How important is simultaneous adaptation to a variety of selection factors in maintaining population genetic diversity?

Sustainable agriculture research must also reach beyond the plant sciences. In the development of seed crop polycultures modeled on the prairie, it will also be crucial to imitate in agriculture the underground behavior of the prairie ecosystem. Unfortunately, the largest knowledge gap by far is what happens below ground. Soil scientists will therefore be needed to help understand the factors involved in maintaining soil fertility within perennial systems. They will help create an appreciation of the workings of the soil's biological community so that growers can work with it rather than against it. Scientists must better elucidate the interactions among climate, vegetation, and soil organisms that affect decomposition, mineralization, uptake, and nutrient transfer between species. Basic studies on root competition are needed to predict how two or more species with different root architectures can better utilize limited supplies of water and nutrients. Other issues for soil scientists to explore include how to measure soil health; how to understand soil genesis as a process of interrelated and interdependent ecological, biochemical, geological, and sociological factors; how to develop unique, appropriate management programs and cropping patterns for particular soils; understanding the functions of different groups of soil organisms; how various soil management practices affect the various groups/species of soil organisms and how to support beneficial groups; and how climate and cropping practices affect soil and soil processes.

Entomologists can assist in the understanding of the modes of plant resistance to insects, whether they be chemical or morphological defenses. Insect ecologists can also contribute to an understanding of the basic biology of problem species and to investigations into how cropping design affects herbivore population density and levels of predators and parasitoids. Research that explores the dynamics of natural enemies and their hosts or prey, or identifies the characteristics of these organisms that lead to economic pest control, will aid in developing successful perennial polycultures. Theoretical work in behavioral ecology and foraging theory will enable researchers to define the attributes of effective natural enemies and the habitats that promote them. When applied to the practical challenge of designing cropping systems, this knowledge will help limit the number of choices of potential arrangements to be tested.

As a final example, animal scientists can study the feasibility of incorporating large livestock, as analogs of wild bison perhaps, into perennial polyculture. As researchers move to consider the whole farm, they will need to appreciate the potential roles of livestock in the fertility and management of perennial plots. The incorporation of these animals is important since large grazers were crucial historically to the development of the prairie ecosystem.

Beyond the need for multidisciplinary scientific research, the development of sustainable agriculture clearly requires that biologists, social scientists, farmers, and many others work together. Taking all of their perspectives into consideration is essential to answering some crucial questions. What are the root causes of the problem? What are the short- and long-term costs and benefits of the proposed solutions? What time and resources are required to effect the proposed solutions? How will the recommended innovations be adopted by farmers? What are the risks associated with their adoption? What are the local and regional impacts of adoption? Those working toward solutions for agricultural sustainability must recognize that the practitioner has at least as much to contribute to the process of understanding biological systems as does the investigator. Sociocultural, economic, and political factors all must be considered as part of any investigation to obtain a more complete understanding of why certain agricultural practices work and others do not. After all, and as discussed earlier in this book, important agricultural factors (soil quality, for example) are indirectly, but dramatically, affected by sociocultural, economic, as well as political factors.

Clearly, the long-term sustainability of agriculture, in the face of dwindling resources and ongoing environmental damage, will depend upon creative, multidisciplinary approaches that combine ecological theories with practical agricultural research to reduce fossil fuel dependency and pollution while maintaining acceptable levels of production and enhancing soil fertility. The blend of ecology and agriculture proposed here could provide for more environmentally benign food and fiber production. It could also broaden the justification for preserving pristine habitats and their species, as these become the ecological standards for agricultural sustainability. The time is ripe for a blend of ecology and agricultural sciences and for a humble search into natural ecosystems for the patterns and properties transferable to sustainable forms of agriculture. Some of the scientific, economic, and societal implications of such a shift form the subject of the next, and final, chapter.

6

Culturing Sustainability

THE ASILOMAR DECLARATION for Sustainable Agriculture states: "A sustainable agriculture will require and support a sustainable society. Our challenge is to meet human needs without denying our descendants' birthright to the natural inheritance of this planet."[1] While farms designed in nature's image may provide the ecological ingredients of sustainability, this is only part of the picture of sustainable agriculture, for it is also necessary to address how people fit into the system. The scientific feasibility of perennial polycultures may be irrelevant if they prove to be impractical from a social, economic, or institutional viewpoint.

This chapter addresses some of the human aspects of the feasibility of perennial polycultures. Whereas some aspects of present-day culture are highly conservative and bode ill for such a radical change in how food is grown, other signs are hopeful, indicating that change is already afoot. While a thorough review of all the consequences of such a change is beyond the authors' predictive powers, the chapter begins by addressing three important concerns: food costs, acceptance of new foods, and effects on farm size, profitability, and social sustainability. The rest of the chapter focuses on preconditions that are necessary to make a change to perennial polyculture farming feasible in human terms.

197

SOME PRACTICAL CONSIDERATIONS

One important practical consideration is how much food grown in perennial polycultures would cost. Clearly, the answer is unknowable until the crops are developed, as well as the cropping systems. At this point, one can only speculate by considering what goes into current food costs, component by component.

One important factor is the cost of input products that are highly dependent on the cost of energy. This cost will surely decline if perennial polycultures can maintain their own fertility, or even partly substitute for external fertilizer inputs, and manage pests without chemicals. Seed costs will also decline if fields are not replanted each year, depending on the cost of seed and seeding rate. Likewise, the cost and number of machinery operations will decline with less soil preparation, planting, and cultivation required in perennial systems. Perennial polycultures might even make the use of draft animals look attractive again. On the other hand, harvesting may become more complex with several crops in one field. How this would translate into costs is unclear. With wide adoption of perennial polycultures and with land divided among several crops, it is possible that the vast, price-suppressing surplus production of corn, wheat, and other crops now prevalent in the United States would become less common. Overall, it seems probable that food production costs would decline with perennial polycultures, but it is unclear how lower costs would translate to consumer prices.

A second practical consideration of a shift to perennial polycultures is whether people would accept the new grains as food. This is also difficult to answer before the crops themselves are actually developed. The American public does seem amenable to new foods, especially to new styles of preparing food, but has not experienced a major substitution of staple grains in recent history. Some other parts of the world have experienced this substitution in the past few decades, however, as the products and processes of modern agriculture were exported around the globe. Food aid also brought new foods, especially wheat and maize, into parts of the world where they were not previously part of the diet.[2] Zimbabwe experienced a shift in staple grains from traditional sorghum and millet to maize beginning in the 1920s and 1930s, when maize was introduced by the British on large, commercial, British-owned farms. By the early 1960s, maize constituted 45 percent of the average caloric intake for Zimbabweans, and by the mid-1970s the percentage had risen to 55 percent.[3] Although the social and ecological consequences of this kind of switch often have been negative, the fact that major dietary shifts have occurred indicates a degree of human dietary flexibility. Nicaragua provides another example where a population has shifted its primary grains, in this case from maize to wheat and back again to maize. Here, the later shift was deliberate, and necessary, in order to achieve food self-sufficiency after the United States cut off wheat exports to Nicaragua in the 1980s. In *Nicaragua: What Difference Could a Revolution Make?* Joseph Collins reported that public acceptance of maize was achieved by an advertising campaign that appealed to national pride. These examples suggest that food preferences are neither rigid nor unchangeable and that new grains can be accepted.[4]

However, the goal is not a total replacement of conventional crops with new perennial crops. One of the beauties of perennial polyculture is its suitability for marginal lands that could not otherwise be cropped sustainably, leaving prime, flat land for annual crops. It is even possible to grow corn sustainably, though not continuously, on good, flat land. Systems that employ cover crops, ridge tillage, and rotations have been developed by farmers who shun chemicals. These practices are used by an increasing number of farmers, including Sharon and Dick Thompson of Boone, Iowa, and by researchers at the Rodale Research Center.[5] These practices may be adequate where land is least susceptible to erosion. Also, much of the current use of corn is for animal feed, oil, and processed foods, where substitutions of new grains would be relatively easy. Replacement of wheat products is harder to imagine because the gluten in wheat is necessary for leavened products. One perennial grain, intermediate wheatgrass, does have gluten and has great potential as a partial wheat substitute. It is also possible that annuals could be incorporated into perennial polycultures periodically as first-year cover crops while the perennials get established. How different would these new grains be as food sources? Gamagrass seed bakes and tastes similar to corn in a cornbread recipe, but less is known about the baking and cooking properties or flavor of the other crops under development for perennial polyculture.

A third consideration in whether a shift to perennial polycultures becomes socially and economically feasible is how such a shift would influence the size and number of farms. Would it change agriculture's relationship to nonagricultural society? This book began with a discussion of some of the connections between economic and ecological problems in agriculture. Clearly, the same practices that cause ecological problems have also produced the economic problems that have led to rural depopulation and lack of social sustainability in many farm regions. Are perennial polycultures compatible with a return to a pattern of more farmers as owner-operators, on smaller farms, with a revitalized rural economic base?

Chuck Hassebrook of the Center for Rural Affairs points to three goals for ecologically sustainable practice that also promote aspects of social sustainability. Although the analysis is necessarily speculative, these three goals can be used to measure the potential of perennial polyculture for promoting social sustainability.

First, Hassebrook suggests that practices that reduce purchases of farm inputs from outside sources will have more social benefits than those that merely shift from purchased petrochemical-based inputs to purchased "environmentally safe" products. Lower costs of purchased inputs reduce barriers to young farmers entering the field and even out the advantage of volume purchase now held by large operations. Fertilizer and pesticide costs are currently the largest share of the variable costs in nonirrigated farming (expenses that don't include such fixed expenses as mortgage payments) for many of the major crops. For corn, fertilizer and pesticide costs represented about 55 percent of the variable costs, or 34 percent of total costs in 1986. For wheat, the figures were 40 percent and 23 percent, respectively.[6] Perennial polycultures are designed to provide much of their own fertility and internal pest control and therefore should require fewer purchased inputs. Also, less seed is required for perennial crops because they are not planted annually. Annual seed purchases constituted bout 15 percent of the variable costs of corn production and 12 percent for wheat production in 1986, according to USDA figures. Thus perennial polycultures meet this first criterion of compatibility with the goal of social sustainability.

The second goal for a socially sustainable agriculture is to increase the rewards for farmers' skilled labor and management rather than increase the need for much unskilled labor or complex investment skills. In other words, we want to avoid solutions that depend on large numbers of unskilled laborers.[7] How does perennial polyculture help achieve this goal? While this system may reduce some of the labor involved in soil preparation and planting each year, it will likely require more attention from the farmer. Farmers will have to pay attention to how their mixtures are doing in different fields and parts of fields and make adjustments if necessary. Shifts in the predominance of the different species in a mixture in different portions of fields will indicate differences in microhabitat, such as drier, less fertile soils, etc. An alert farmer can adjust planting ratios or manage fields to increase yield, but it will take greater attentiveness, not just more chemical applications. The farmer will also need to pay attention to potential pests and work with the biology of the system to minimize pests' negative impacts. This extra vigilance may mean that farm success will be related to farm size in a new way. It may be critical that a perennial polyculture farm not be so big that the farmer can't attend to it all. Farm size may become limited not by the size and speed of machinery, but by the attentiveness, skill, experience, and hands-on management of the farmer and helpers. As Wes Jackson puts it, the critical factor determining farm size may become the "eyes-to-acres ratio." Thus, perennial polyculture agriculture can be expected to promote the goal of rewarding farmers more for sharp wits and creative thinking and less for drudgery and routine heavy labor.

A third goal for socially sustainable agriculture is that it can be applied at least as economically on moderate-size farms as on large farms. In addition to the possible attentiveness benefits (noted above) of moderate-size compared to large perennial polyculture farms, it can be anticipated that these systems would be less risky than single-product farms. Multiple crops or enterprises on a farm reduce the economic risks for farmers.[8] Perennial polycultures may do this even better than rotations of annuals grown in monocultures. As discussed in Chapter 5, polycultures have the potential advantage of overyielding. This reduced risk is available to small farms as well as large farms. It also improves the odds for success for new farmers as they begin farming.[9]

On the other hand, perennial polyculture would require considerably less labor than a diversified organic farm using annual crops, for instance. There could be room for fairly large farms also, but these would not necessarily have the advantages of the economies of scale that large growers now have over smaller growers. The timing of labor would certainly be different. Less spring work would be expected because no planting is necessary most years. This also relieves the risk of poor weather at planting time. Risks of crop failure or soil compaction from working soggy fields would be lower for farmers of perennial crops, regardless of farm size. Thus, it can be speculated that large perennial polyculture farms would not have appreciable advantages over smaller farms in the arenas of inputs, on-farm expenses, and labor efficiency, just by virtue of their size. However, in the arena of marketing, perennial polycultures would not alter the advantage of large size. This issue requires the attention of policymakers.

Perennial polycultures also offer the potential to combine seed production with livestock grazing in the same fields in different seasons. This is a natural combination for species that evolved with large grazers, such as most prairie species. This option would further promote on-farm diversity and could add to the financial stability of farms.

Finally, one more aspect of social sustainability may be affected by a conversion to perennial polycultures—a sense of rural community. Two aspects of perennial polyculture farming are likely to promote interactions among neighboring farms. For one, dependence on natural biological dynamics for pest control would mean that practices on one farm would affect those on a neighbor's farm. Second, the need to adjust a perennial crop mixture to local conditions would mean that a farmer's best source of information would be his or her next door neighbor rather than an agricultural experiment station in another county. These interdependencies could foster a cooperative atmosphere and strong sense of community in rural areas.

INCENTIVES FOR CHANGE

A shift to perennial polycultures requires shifts in the underlying goal of agricultural improvement such that long-term preservation of a resource base that includes farmland, soil, and farmers is valued above potential individual gain. To achieve this will require a coordinated group effort to reinforce common goals. At the 1990 Ecological Farming Conference in Asilomar, California, seven challenges were outlined as necessary steps to facilitate a switch to sustainable agriculture. The Asilomar Declaration lists these challenges:

• Promote and sustain healthy rural communities.
• Expand opportunities for new and existing farmers to prosper using sustainable systems.
• Inspire the public to value safe and healthful food.

- Foster an ethic of land stewardship and humaneness in the treatment of farm animals.
- Expand knowledge and access to information about sustainable agriculture.
- Reform the relationship among government, industry, and agriculture.
- Redefine the role of U.S. agriculture in the global community.

While a detailed discussion of policies and programs to respond to these challenges is beyond the scope of this book, it is clear that the success of perennial polyculture farming will depend on long-term tenure of the land. Opportunities for new farmers, and enough people on the land to support viable rural communities and the support systems that add to the quality of rural life, are also essential for any form of agriculture that proposes to endure. It is also essential that farmers are financially rewarded for preserving the soil, minimizing environmental impacts, and maintaining healthy agroecosystems, not for depleting the soil and polluting the environment.[10]

The 1985 and 1990 Farm Bills took a few steps in the direction of linking soil conservation practices with farm support benefits. In the 1985 bill, the Conservation Reserve Program (CRP) allowed farmers to take highly erodible land out of production for ten years in exchange for a rent payment from the government. The land had to be put under permanent cover, and the acreage set aside this way was removed from the farm's base acreage eligible for commodity subsidies. It was thus a dual-purpose program designed to protect erodible land and to reduce crop surpluses. The CRP was the "carrot" portion of the bill. It also had a "stick" that required farmers with highly erodible land to design approved conservation plans by 1990 to remain eligible for any government farm support or loan program. The catch in this bill was that other provisions still made farmers lose benefits if they used crop rotations. In the 1990 bill, a step was taken toward facilitating farmers' use of crop rotations. Rotations have the potential to supply soil nitrogen and to control weeds and other pests without chemicals. The potential benefit of this provision, which allows farmers to maintain their base acreage for commodity programs when using rotations, may be offset by across-the-board cuts in commodity payments that disadvantage smaller producers much more than larger producers.[11] Farm policy is changing, but it still has a long way to go before it will support truly sustainable agricultural practices.

Intermediate steps between the ultimate ecological sustainability of agriculture and the system currently in place will be essential. Some signs indicate that social value shifts are already in progress. For example, the 1990 Farm Bill provided the first national standards for "organically grown" labeling. As society begins to value agricultural methods that promote the health of children, workers, and the environment, organic produce is showing up in major grocery chains and selling at higher prices than conventionally grown food. Paying premium prices for organic products is one way to support farmers as they make the transition to methods of production that seem riskier financially. Society is apparently willing to assume at least some portion of the cost, but this higher cost excludes from the benefits those who cannot afford higher-priced food. This dilemma needs to be addressed directly, for the "cheap food" policy of the United States is one impediment to adoption of organic practices. Typically, Americans spend 11 percent of their income on food, the lowest percentage in the developed world. The French spend 17 percent of their income on food; the Japanese, 20 percent; the Greeks, 37 percent. Still, even with low-cost food, hunger is a serious problem in the United States. Thus, although low food prices (as Noel Perrin noted in "The High Price of Agribusiness") trap farmers into a high-input, high-output, high-pollution, high-waste system just to survive, higher prices for farm products is a difficult policy to contemplate. Nevertheless, it may be necessary in the future to pay prices for food that more realistically reflect the costs to farmers of taking on the risks of a shift to more sustainable methods. If so, it will also be essential to devise plans to make an adequate diet available to all citizens. Further, as noted in Chapter 4, simply going "organic" may not preserve precious soil and oil resources in some cases. To achieve the more radical changes necessary for truly sustainable grain production, even greater shifts in the social, political, and economic arenas will have to occur.

DRAWING ON FORCES FOR CHANGE

The ideas presented in these pages are products of our current culture, which is itself in the process of change. A variety of groups are working from different angles to build a more sustainable agriculture within a more sustainable society. These changes must be nurtured along with scientific research, if the changes suggested here are ever to come about.

GRASSROOTS CHANGES

One example of a group working toward agricultural change from a different angle is the Land Stewardship Project in Marine, Minnesota. This group seeks to foster and nourish a land stewardship ethic among farmers. Many farmers have a profound desire to be good stewards of their land but have felt trapped by economics into using poor stewardship practices. By supporting one another's efforts, and even going after some sources of problems (for example, changing local ordinances to compel farmers to employ soil conservation plans in cases where erosion is affecting a neighbor's land, and fighting for stewardship standards for corporate owners in their Farmland Investor Accountability Program), this group is creating ripples of change.[12]

Another concept new to the United States is community-supported agriculture (CSA), or subscription farming, wherein consumers share the risks of farming by buying shares of a farmer's crop at the beginning of the season. Thus, before planting, the farmer knows the size of the expected market and has a guaranteed income. In return, shareholders receive a portion of the harvest of fresh foods throughout the season. The desire to know where one's food is coming from, how it is grown, and how fresh it is, as well as a desire to support a farmer directly, are the motivations for joining a CSA.[13] This innovation meets the needs of both the farmer and the consumer and ensures that food is consumed close to where it is grown. The latter is an important point, since transportation is a major cost in our total food system.

ON-FARM RESEARCH

Another ongoing and promising change is a new emphasis on on-farm research. Whereas the traditional form of on-farm "research" has consisted of field trials of new varieties of chemicals that were developed at experiment stations or in private research units, a new wave of on-farm research involves farmers in experimental comparisons, realistic analysis of methods, and other phases of research. This shift promises more farmer input into the nature of the questions asked in agricultural research, more opportunity for individually tailored farming solutions, and new opportunities for continuity in long-term research projects. Farmers expect to have long-term tenure of their land, and they have no pressures on them to publish, as do academic researchers.[14] They are thus ideal candidates for long-term research projects. Moreover, the farmer's livelihood is on the line. Economic and practical questions, as well as mistakes, are not merely academic. In the short term, the farmer's stake in research is fundamentally different from that of the scientist.

Ken McNamara, in an article titled "Research Update: The Big Difference Was Cost," reported that Rodale's Regenerative Agriculture Association's Midwest on-farm research program had thirteen farmer-researchers in midwestern states in 1988. Richard Ness described the Land Stewardship Project's (LSP) Stewardship Farming Program, which promotes and assists on-farm research in "A Paradigm Shift for Extension: On-farm Research in Sustainable Farming Practices." The LSP staff see their roles as facilitators working with farmers. The farmers provide the leadership, and the staff assists in areas such as experimental design and interpretation of results, and also by providing ways for farmers to share their knowledge with one another, or to gain more knowledge about a topic of interest. To this end, they publish a directory and a newsletter and organize farm tours and speakers.

The philosophy of this type of farmer-directed research is that sustainability is a property that must be discovered on each farm, not something that can be dictated from a distant research institution or university, or achieved by recipes and formulas meant for broad application. In this view, sustainability depends on the unique characteristics of each farm: the soil, the history of human use, the farmers' needs, talents, and resources, the neighbors' crops and practices, and the community's needs and resources. Whereas State Agricultural Experiment Stations traditionally have focused on regional problems and tried to tailor agriculture to the special conditions in each state, individual farmers can take this to an even more local and personalized level. On-farm research has been found to be an important element in native cultures where agriculture is still practiced in traditional forms. Miguel Altieri offers examples of how farmers' experiments with different methods through the centuries have resulted in forms of agriculture that suit local conditions and supply food in a sustainable fashion.[15] Certainly, having farmers involved in trying to solve their own problems will heighten farmers' attentiveness, increase the likelihood that problems will be detected, and improve the "eyes-to-acres ratio" discussed earlier. The Stewardship Farming Program invites participation by "any farm families interested in incorporating their concerns for water, soil, health, economic sustainability and the viability of their community into their farming practices."[16]

Dick and Sharon Thompson of Boone County, Iowa, espouse a similar philosophy. They stopped using chemicals in 1967 and began experimenting with different farming methods to make nonchemical farming work for them. In his article "The World Discovers Nature's Ag School," Craig Cramer described how the Thompsons recently joined forces with the Rodale Institute to develop scientifically valid experimental designs. The Thompsons have developed a large network of farmers in Iowa who conduct their own research to answer their own questions about what will work to make individual farms more sustainable. The group is called Practical Farmers of Iowa. The Thompsons also collaborate with land grant and USDA scientists on other research projects, including such things as studying whether manuring soybeans leads to groundwater contamination. This sort of collaboration is especially worthwhile for projects that require expertise in special techniques or special analytical equipment.

The way that the Stewardship Farming Program (SFP) and Practical Farmers of Iowa (PFI) operate is in sharp contrast to the traditional role of the Cooperative Extension Services, where results from research conducted at university and USDA laboratories are handed down to farmers. Both SFP and PFI work from the bottom up, in that farmers are the experts. Farmers generate the research questions and help each other solve them. Scientists act as technical consultants in these programs. Organizations such as SFP and Rodale's national Farmers' Own Network for Education seek to match up farmers who have questions with farmers who have pertinent experience. These organizations use a horizontal network structure rather than the traditional hierarchical structure of the university extension system. A combination of more on-farm research and farmer-to-farmer extension networks has great potential for speeding up the development and transition to sustainable agriculture.

Much basic research—such as understanding the ecological relations that make natural ecosystems sustainable or better understanding the biological dynamics of soil communities—may always be conducted best at research stations by trained scientists. Still, much applied research may be better conducted on farms, or at least initiated from forums that allow for farmer input in identifying important questions.

Among the advantages of on-farm research into practical problems are:

- a better match of agricultural practices to local conditions, such as soil type
- larger tracts than are available at field stations
- the possibility of whole-farm or larger-scale studies that encompass a more realistic range of growing conditions
- more realism in terms of how practices or crop varieties function under real farm conditions and within the constraints of other operations, farmer's skills, etc.[17]

On-farm research is attracting the attention of traditional agricultural research institutions. For example, Dennis Keeney, a researcher at Iowa State University's Leopold Center for Sustainable Agriculture, has called for more farmer participation in research. A conference on "Farmers' Role in Farming Systems Research and Extension" in the fall of 1990 was sponsored by another land grant university, Michigan State University. The USDA's Low-Input Sustainable Agriculture (LISA) competitive grants program funded some 90 projects in its first three years (1988–90) that involved more than 700 farmers in a variety of roles, from helping plan and evaluate research to sharing information with other farmers. The 1990 Farm Bill included a $20 million provision for a special outreach program in sustainable agriculture, which, in addition to providing for extension agent training in sustainable agriculture, includes assistance for farmer-to-farmer networks and input into the USDA research agenda.

CONSERVATION, PRESERVATION, AND AGRICULTURE AS ALLIES

Although preservation and conservation have long been at odds, the two are beginning to converge of late. Preservationists are those who wish to preserve nature intact, for its own sake. They are represented by such groups as the National Audubon Society, the Wilderness Society, and The Nature Conservancy. Conservationists, on the other hand, are devoted to the wise use of natural resources and sustainable development, and are represented by groups concerned with wildlife and fisheries, forestry, and soil conservation. While preservation has often been at odds with agriculture—for example, regarding deforestation for agricultural development, the draining of wetlands to create new fields, and agricultural pesticide effects on wildlife—conservation and agriculture have had a close history, with soil conservation being the most obvious agricultural manifestation of this relationship. The boundaries between preservationists and conservationists have blurred in the past, however, and are blurring more and more at present, as many people come to understand that human activities are based in nature and that all of nature responds to human activities. Interestingly, the prophet of preservation, the eloquent spokesman for a land ethic, Aldo Leopold, started as a conservationist, a wildlife biologist whose first writings were calls for antipredator management techniques. The late convergence of these movements is not spurious; it is clearly coming about because of the critical effects the human population has had on the earth. Human impacts are now threatening the last natural ecosystems and wilderness on the earth.

The preservation movement is a crucial ally for sustainable agriculture, and vice versa. In fact, one depends on the other. Both preservation and conservation demand a sustainable agriculture. For if there are to remain any natural habitats left to preserve, or natural resources to conserve, then we must be able to contain our agricultural land close to its current bounds without plowing up any more wetlands or forests. We must be able to ensure that we can continue in this manner indefinitely, that our agricultural land will remain sufficiently productive indefinitely to accommodate human food needs. Also, we must minimize negative impacts of agriculture on natural ecosystems. In particular, this requires less run-off and leaching of agricultural chemicals, less deposition of eroded soil in aquatic systems, less wholesale alteration of natural lands for agricultural purposes, and less demand for petroleum-based products and water for agricultural lands.

Similarly, agriculturalists need preserved natural habitat. It is important to consider ecosystem, even ecosphere,[18] function in plans for sustainable agriculture. Natural ecosystems and their suite of wild species are an important resource for sustainable agriculture. Here dwell the sources of genetic diversity necessary to obtain adaptations of crops to various environments, or resistance to new pests, as well as many beneficial organisms to aid in pest control and nutrient recycling. Here also are the standards of ecosystem health, the source of knowledge about what works in ecosystems, the model to mimic in managed agricultural systems.

Yet so much of the landscape has been altered by human use that only small remnants remain as standards for sustainable agriculture. In Iowa, less than 0.2 percent of the original prairie remains, and the last major native grassland stand is no more than 200 acres.[19] The situation is similar in Illinois, where only about 2100 acres of high-quality prairie remain, entirely in small tracts. In the early 1800s, prairie accounted for more than 60 percent of the Illinois landscape.[20] Natural ecosystems may also serve as valuable buffer zones that prevent epidemic spreads of pests, store water, process wastes, and even moderate climate. Natural ecosystems perform such vital global stabilizing processes as holding water, consuming carbon dioxide, and filtering and cleaning water.

Furthermore, both preservationists and sustainable agriculturalists must come to recognize that humans truly are part of nature too. Both movements must consider human use within the context of ecological principles in order to achieve their respective goals. For conservationists, it is becoming more important to include plans for meeting human needs along with preservation plans in order to securely protect natural habitats. And it is important to meet these needs in a sustainable fashion. In Latin America, where human needs intensely pressure natural habitats, this is particularly obvious.[21] Brazil is establishing "extractive reserves," which are essentially intact forests reserved for use by rubber tappers, Brazil-nut gatherers, and others who gather products from the forest without destroying the plants, but not for agriculture or cutting.[22] This type of preserve is especially important because much can be learned from native cultures that have integrated human use into natural ecosystems for many generations. Preserve planners and agriculturalists alike would do well to tap such folk practices while they still exist.

The Nature Conservancy's (TNC) experience in Latin America is now being applied in the United States. Curtis Badger, in an article titled "Eastern Shore Gold," described how at TNC's Virginia Coast Reserve the concept of integrating preservation and human use is becoming a reality. If, for example, the coastal islands of Virginia are to be preserved, the compatible human activities of farming and fishing on the adjacent mainland must be preserved and not given over to intensive housing developments. Agriculture must be practiced so as not to impact fisheries, and agricultural land must be protected from being taxed at development potential rates. Similarly, if conservationists wish to preserve old-growth forests in the Pacific Northwest, provision for the livelihoods of loggers must be made—a sustainable forestry and a sustainable level of demand for forest products must be developed.

Thus it appears that agriculture, conservation, and preservation are moving toward common ground. It is encouraging that organizations traditionally most concerned with the preservation of pristine nature are beginning to consider agricultural issues also. *Audubon* magazine published an issue in November 1989 that featured five articles on agriculture. The new strategic vision for the 1990s at The Nature Conservancy is creation of a number of "bioreserves," large segments of "critically important ecosystems" that include important pristine habitats and also human-use areas.[23] The developing alliance of sustainable agriculture with conservation and preservation will speed the progress toward these movements' interrelated goals.

CHANGES AND CHALLENGES FOR AGRICULTURAL RESEARCH

At the same time that grassroots pressure is mounting for a more ecological and humane agriculture, with the growth of organic-certification movements, low-input land stewardship and field worker movements, rural revitalization and consumer awareness movements, and farmer-to-farmer extension networks, most agricultural research institutions are veering in the opposite direction in their research agendas. Throughout this book it has been emphasized that the problems that make agriculture nonsustainable today are deeply embedded within modern agriculture, from research institution to farm to market, and that a shift to sustainable agriculture requires some fundamentally different ways of approaching these problems. It has been argued that a sustainable agriculture that can function without massive fossil-fuel energy subsidies, provide much of its own fertility, protect and build soil, and minimize the effects of harmful organisms will likely resemble native ecosystems more closely than modern monocultures. To achieve these new forms of agriculture, then, it is necessary that research go beyond the current approach of merely tinkering with monocultures and into new territory of assembling agricultural analogs of natural ecosystems. It is going to take a much more extensive research effort than can be maintained by the few, small, mostly private programs now involved. The tremendous potential of the already established land grant, SAES, and USDA research network must be tapped, as well as the growing network of farmer-researchers.

Although interest is increasing—seventeen land grant universities now claim some sort of sustainable agriculture program[24]—it is not clear that the traditional research institutions are able or committed to take up the search for a truly sustainable, sunshine-powered agriculture. A number of barriers exist and must be challenged to reach this goal: barriers that prevent interdisciplinary studies, barriers to long-term research projects, competition for funding (particularly lopsided competition with biotechnology), and conflicting value systems influencing choice of research projects and approaches to problem-solving.

BARRIERS TO INTERDISCIPLINARY WORK

The development of a permanent agriculture demands interdisciplinary efforts. It is essential for ecologists to combine their knowledge of natural ecosystems with agriculturalists' practical knowledge of crop breeding and animal science. Yet this is a major challenge within the structure of current research support and university systems. These fields are in separate colleges and therefore under different administrative units within universities. This may not be entirely disadvantageous, as at least faculty are not competing with one another for prestige within a department. The complications of working together across departments is made more pronounced by simple physical isolation and lack of name recognition across fields. Even grants to support basic research in these two areas traditionally come from different agencies—the National Science Foundation (NSF) for ecology and the USDA for agriculture.

Interdisciplinary work is comparable to interlingual communication, as different disciplines develop not only their own jargon, but conventions about what constitutes truth, what makes an authority credible, even how to present material publicly. For scientists, a vast store of knowledge of (scientific) facts demonstrates authority, whereas social scientists learn that (social) facts are a human invention. Scientists peak early, and older scientists are often considered deadwood by colleagues. Those in the humanities, on the other hand, are considered to peak late and are venerated by their peers toward the end of their careers. In the humanities, respect is accorded to those with striking originality of thought or expression, rather than those who can provide recitations of extraordinary facts. All these intradisciplinary conventions influence how practitioners approach problems, their choice of research goals, and their satisfaction with different outcomes. Thus, it can be difficult to make progress in interdisciplinary efforts.[25]

Bruce Maxwell, a weed ecologist in the Department of Agronomy and Plant Genetics at the University of Minnesota, has been involved with interdisciplinary groups including social sciences, ecology, and agricultural disciplines in Minnesota and Oregon.[26] His experience confirms that it takes a great deal of patience to make progress in interdisciplinary groups, and the fact that most of those involved are young, untenured faculty in need of publications to show for their efforts means that patience runs thin quickly. Maxwell found that in both groups participants spent the first several meetings going over the same ground repeatedly until everyone could understand each other's language and approach. He suggests that one way to facilitate more interdisciplinary efforts in the future is to build cross-disciplinary study into graduate student training. It also helps to have one member of an interdisciplinary group who has training in group dynamics. Such training might also be included in graduate work.

Mixing disciplines within the sciences is apparently much less difficult than mixing science with the humanities. Phil Robertson, one of the principal investigators on Michigan State University's Agro-ecology Long-Term Ecological Research project, has found that the seventeen faculty from seven different sciences involved in this project have communicated quite well.[27] This group functions as a collection of semi-autonomous investigators working on many different projects that all fall under the umbrella of a global hypothesis. The hypothesis is broad enough to encompass many projects and general enough that it allows autonomy. The group's purpose is to discover basic relationships and ecological concepts that can substitute for chemical subsidies in production-level agriculture. So far the group is not aiming for a group product, but does see extension applications eventually coming out of the research. Farmers have participated in the original choice of cropping systems compared on the research site and will be more involved as extension applications develop.

Some loss of personal autonomy is usually necessary for interdisciplinary work, and research institutions will have to accommodate this collectivism if sustainable agriculture research is to progress. This accommodation process will involve working out ways to shift the basis of rewards from the current emphasis on autonomy and individual achievement to also include focused, group efforts when such decisions as tenure, promotions, salary, and discretionary funding grants are made. It is not clear at this time whether or not any such shift is already in progress, or even desired by university administrators.

One way around the difficulties of cross-disciplinary research collaboration is to start a whole new field that develops its own literature and approach to problem-solving and eventually spins off its own specialized subdisciplines. To some extent this process is already occurring. The field of agroecology is growing rapidly. Stephen R. Gliessman, in his essay "An Agroecological Approach to Agriculture," defined the field as the science of ecology applied to solving agricultural production problems. Although this definition suggests a problem-solving orientation, to date much of the research in agroecology has been more exploratory than problem-oriented. Agroecology must not get bogged down in endless exploration, for this will only slow progress toward sustainability. What is needed is a more focused problem-solving mode, deliberately combining science and the humanities and expressly designed to revise and reshape agriculture according to an ecologically sound model, a model to be sought in natural ecosystems.

BARRIERS TO LONG-TERM RESEARCH

Another barrier to sustainable agriculture research in traditional institutions is the necessarily long-term nature of such work. State Agricultural Experiment Stations have traditionally been funded primarily from state appropriations supported by state constituencies. They have thus been responsive to local concerns, and research has been directed to solve immediate, pressing problems. This funding mechanism has not been conducive to long-run solutions. It has reinforced the tendency to go with chemicals, rather than IPM, for instance.[28] Lately, pressure for more basic research has resulted in more federal competitive grant money available for agricultural research. But these grants are also short-term (two to three years) and thus do not facilitate long-term projects.

The entire process of university employment is also based on rapid research results. Published papers on research results must be amassed quickly for an impressive curriculum vitae, and success often entails several moves over the course of a career. Neither are conducive to long-term research projects. Thus new ways of funding long-term projects are needed, as well as ways to support permanence in research. In fact, there is a call now among many scientists to increase NSF grant award periods from three years to six years.[29] A shift in the current standards for tenure and promotion is also necessary to facilitate this kind of research. The prospects for such a shift are not particularly promising at this point. The ground swell for sustainable alternatives will probably have to grow larger and louder before traditional research institutions make a substantial response.

COMPETITION WITH BIOTECHNOLOGY FOR FUNDING

Perhaps the major hindrance to a concentrated effort to develop more sustainable agriculture research programs is competition for funding. Many agricultural research institutions are moving in the opposite direction from the holistic, whole-systems approaches of sustainability studies, toward the opposite end of the spectrum of biological organization: they are instead investing in molecular biology and genetic engineering, the biotechnology approach. In some cases, a particular institution is exploring both approaches. For example, Michigan State University recently filled the endowed Mott Chair in Sustainable Agriculture with Richard Harwood, a bold choice that speaks strongly of a serious commitment to developing sustainable agriculture. Harwood served on the National Academy of Science task force that produced *Alternative Agriculture* in 1990. He worked at Winrock International Institute of Agricultural Development for six years and directed the Rodale Research Center from 1977 to 1984. At the same time, MSU remains closely involved with biotechnology research. The Michigan Biotechnology Institute (MBI) is an industry liaison group within the university that seeks funding and marketing opportunities for university-based biotechnology research. In a move that stretched the ideological separation of publicly funded research from for-profit enterprises, MSU created a private for-profit corporation, Neogen, to develop and market products of their biotechnology research.[30] Yet what may appear to be a solid dual commitment on the part of the university is not quite so. The sustainable agriculture program is largely sponsored by private foundation funds. The Charles Stuart Mott foundation funded the endowed chair and four graduate student fellowships. Hopefully these foundation grants will serve as seed money to get a more extensive program started, but it remains to be seen what hard choices will be made when biotechnology and sustainable agriculture do have to compete for support at MSU. Frederick Buttel of Cornell University, a frequent spokesman on sustainable agriculture issues, predicts in his chapter in *Agroecology* that this combination—biotechnology and alternative agriculture—may become more common at State Agricultural Experiment Stations. Alternative agriculture research will act to expand the political base of support while biotechnology research maintains a grip on industrial funding sources.

Despite many hopeful commentaries on the potential for biotechnology to aid in achieving a sustainable agriculture, much evidence points to biotechnology research dollars going toward marketability, not sustainability.[31] Agricultural market potential for biotechnology products is estimated at anywhere from $2 billion to $21 billion. Even the lower estimate represents a large incentive for market-oriented research. The dollar amounts involved in biotechnology research are so huge that research institutions are adjusting their staffing patterns to accommodate this new research direction. By 1986, 430 full-time equivalent faculty positions nationwide were devoted to biotechnology. While 200 molecular biologists were added to the State Agricultural Experiment Stations in the mid-1980s, 115 plant and animal breeder positions were lost.[32] In 1988, the National Research Council's Board on Agriculture recommended a sixfold increase in agricultural biotechnology funding from $84 million in 1988 to $500 million by 1990. In comparison, the 1990 Farm Bill authorized less than one-tenth that amount—$40 million—for sustainable agriculture research in the LISA competitive grants program. This was nearly a tenfold increase over the 1990 LISA program, but only $6.7 million of it was actually appropriated for 1991 (one and a half times the previous year's appropriation). Appropriations for the fiscal year 1991 National Research Initiative competitive grants program included $52 million for "Plant and Animal Systems," categories including biotechnology topics, compared to $13 million for "Natural Resources and Environment," the category where most sustainable agriculture research would find funding. By its highly technical nature, biotechnology research is extremely expensive compared to typical sustainable agriculture research. Because it is such a major consumer of public agricultural research funds, vigorous public debate on biotechnology, its implications, and its funding is called for.

In addition to growing public funding of biotechnology research, private research has expanded even faster. As of 1988, some fifty biotechnology companies in the United States were engaged in agricultural biotechnology applications. These firms were already spending more than $200 million per year on in-house research and additionally awarding $120 million in grants and contracts to universities in 1984.[33] With this strong component of private funding, it is no surprise that within biotechnology research, market potential is a stronger directing force than agricultural sustainability.

Further, since biotechnology research is concentrated in the giant agrichemical firms, it is unlikely that this research will produce alternatives that render agricultural chemicals obsolete.[34] Therefore, the suggestions that biotechnology could be used to transfer nitrogen-fixing potential from legumes to other crops, to reduce dependence on inorganic fertilizers, or to splice disease and insect resistance into crops to reduce the need for pesticides, are unlikely to come out of privately funded research. In fact, a great deal of effort is going into transferring herbicide resistance to crops (such as soybeans) to facilitate wider use of herbicides.[35] Herbicide-resistant crops would be an advantage for no-till farming, but it is doubtful whether wider use of herbicides will prove to be ecologically sound, or even necessary. Some reports now indicate that after the first few years of using no-till methods, weeds naturally decline and herbicides may no longer be needed.[36]

Choices of research direction exist within biotechnology, to be sure, choices for more or less environmentally benign products. But even if research shifted to emphasize pest resistance rather than herbicide resistance, some fundamental incongruities would remain in the concept that biotechnology can be used as a tool to achieve sustainable agriculture.

For one, a number of serious concerns have been brought forth about biotechnology. Some of these center on accidental consequences of introducing genetically engineered organisms into the environment.[37] One concern is that the genes that produce super-crops—those that are resistant to pests, supply their own nitrogen, or are resistant to drought—could be transferred to wild populations. Crops and weeds have been known to hybridize to produce aggressive, weedy crop mimics.[38] The fear and strong possibility are that, in the process of producing super-resistant crops, super-resistant weeds would also be created. Another worry is that the crops themselves would become major weeds, hard to eliminate from fields when a farmer wished to rotate crops, for instance.

The main incongruity between biotechnology and sustainable solutions may not be the accidental consequences, but the deliberate consequences.[39] Biotechnologies aim to further homogenize agriculture. The research is so expensive that a product must have a large market to pay for itself. While this criterion might not exclude sustainability, it is rather unlikely that the two would coincide. As already discussed, sustainable agriculture must be adjustable and variable in differing ecological settings. Biotechnology products are also likely to be more expensive than low-technology counterparts. They will be promoted on the basis of ease of handling—for example, coated seeds that deliver chemicals and seeds in one operation[40]—rather than low cost. But sustainable agriculture will require more modestly sized farms to accommodate the farmer's attention and provide for healthy rural communities—and, therefore, more modest, not higher, input costs. High-priced inputs are aimed at huge producers, who can obtain discounts on input prices because they buy in large quantities. Thus biotechnology will tend to exacerbate the tendency for large farms to benefit more than small farms from new technology and thus will work counter to the goal of sustainability.

The basic philosophy and approach of biotechnology are incongruous with the concept of sustainable agriculture. Sustainable agriculture comes out of the higher orders of organizations, seeking models in ecosystems and working to integrate human use with natural ecosystem functioning. Biotechnology comes from the lower end of the organizational hierarchy, from subcellular, molecular biology. Biotechnology and sustainable agriculture epitomize the reductionist/holistic dichotomy outlined in Chapter 2. This raises the question: By which approach are the solutions to sustainability likely to be found?

Further, biotechnology research is aimed at technologies that in-crease production. These promise to increase surplus production, thereby suppressing farmers' prices and driving more farmers out of business. Another major aim of biotechnology is to transfer produc-tion of plant-derived chemicals (for example, vanilla and rubber) from farm to factory. Many of these plant extracts are produced in developing nations, and the loss of markets for these natural prod-ucts would mean an even greater burden of poverty, with profound implications for Third World debt load and hence for world peace.[41]

At the heart of these problems is the fact that biotechnology pro-vides no new accounting for natural biological processes and rela-tionships, and no new accounting for social-ecological interactions. It is just one more step, albeit perhaps the ultimate step along the path that society has followed to this point, the path toward conquer-ing nature and controlling the environment. Biotechnology is thor-oughly embedded within the scientific-industrial paradigm of "man over nature" that has taken us to the brink of environmental disaster, where we now perch. The mind-set remains that of a battleground between crops and pests, or a factory setting where the goal is maxi-mum production, rather than a search for sets of species and pro-cesses that minimize the likelihood of any one species ever reaching pest status and maximize the reliance on natural interconnections and processes to retain and make use of nutrients, water, and soil.

The research initiatives projected for the State Agricultural Experi-ment Stations in the late 1980s showed little attention to issues of sustainability. Although top priority went to maintain and protect water quality and quantity—certainly a part of the sustainability is-sue—the second priority initiative, biotechnology, was recom-mended for nearly twice the funding level. Similarly, sustaining soil productivity ranked fourth in priority but ninth in funding (see Table 6.1). Clearly, sustainability did not rank high in terms of funding priorities. While the 1990 Farm Bill clearly called for emphasis on sus-tainability, the 1991 Request for Proposals for the agricultural compet-itive grants program did not place special emphasis on sustainability, nor did funding levels reflect such an emphasis.

The public needs to scrutinize and question the ultimate goals and implications of biotechnology research. While some individuals call for public institutions to counter private biotechnology research with funding for biotechnology research more in the public interest, per-haps a better strategy would be to make a major commitment to de-veloping low-cost, sustainable agricultural systems.

TABLE 6.1

1988 Research Initiatives for the State Agricultural Experiment Stations (As Listed by the Experiment Station Committee on Organization and Policy in Cooperation With the Cooperative State Research Service) Ordered by Recommended Funding Needs and Priority Rank

Initiative	Funding needs[a] (in thousands of dollars)	Priority rank
1. Biotechnology	130,075	2
2. Maintain and Protect Water Quality and Quantity	73,665	1
3. Food Processing, Preservation, and Quality Enhancement	60,500	6
4. Animal Health and Disease	60,500	12
5. New and Expanded Uses for Agricultural and Forest Products	52,500	8
6. Interrelationships of Food and the Nutritional and Health Status of People	47,950	10
7. Animal Efficiency in Food Production	45,375	7
8. Sensors and Computing Systems for Food and Agriculture	45,375	17
9. Sustaining Soil Productivity	30,701	4
10. Integrating Agricultural Technologies	27,000	9

[a] A combination of one-time and continuing funds.

Clearly, in the face of the huge amounts of money involved in biotechnology, it would take tremendous courage and vision for a large agricultural research institution to focus on sustainability instead of biotechnology. Thus it is not surprising that much of the ongoing research toward sustainability is occurring on farms (for example, the Thompsons' farmer network and the Land Stewardship Project), by farmers, or at independent research institutes such as The Land Institute and the Rodale Research Center, and much is supported by grassroots movements with private funding. The big institutions are conservative and slow to respond to change. A large reason for this inertia is that most administrators and senior faculty in agriculture received their training in the 1950s and early 1960s, when a growth-based economy seemed more reasonable than it does today. Many are, frankly, unable to perceive what the sustainability movement is saying. Newer faculty often have their hearts in sustainability, but without tenure or prestige, they are typically overwhelmed by the dominant paradigm. In short, change in the traditional research institutions may simply depend on waiting for a new generation of research scientists to ascend the ladders of administrative authority.

CONFLICTING VALUE SYSTEMS IN RESEARCH

Finally, it is important to acknowledge openly that values and beliefs do indeed influence technology and the science that supports and develops that technology. Belief in the ultimate propriety of competition and the correctness of efficiency and profit built the foundations of today's highly productive, labor-efficient, yet unsustainable agriculture. This view seemed to fit in a world of abundant resources and rapid growth. Today, however, humanity is coming up against global limits to growth. Many signs indicate that the planet is rapidly reaching or even exceeding its capacity to sustain the industrial human species. Acid rain, the thinning of the stratospheric ozone layer, soil erosion, an increase of atmospheric carbon dioxide, and many other phenomena indicate that humans are taxing the ecosphere's resiliency and stability. It is becoming evident that we all must learn to live in something like a steady state, an equilibrium on Earth, and reduce practices that tax the planet's recycling and renewal capacities. The agriculture that seemed to fit during our growth phase no longer makes any sense.

Of course, agriculture is only a small contributor to many of these global problems, yet it is a major contributor to some—for example, degraded and depleted water supplies. Moreover, industrialized agriculture is inconsistent with the steady-state model, dependent as it is on wasting soil and water resources, using nonrenewable fossil fuels, and broadcasting toxic wastes.

To acknowledge openly that what is needed is to replace the values and beliefs on which modern agriculture was founded with values and beliefs that support a steady-state relationship between humans and the earth is a major step toward developing sustainable agriculture. And although the word *sustainability* is getting more media attention, it is still often used in the context of growth and increased production. In an article titled "The Ecology of Sustainable Development," William Rees complained that the meaning of the term *sustainable development* is being altered as it is embraced by the political mainstream. Instead of meaning development that recognizes global environmental decay, economic injustice, and limits to material growth and ensures a sustainable environment, it is coming to mean development that raises productivity and consumption standards in developing countries and ensures sustainable material growth.

As limits to growth become clear, the interconnectedness of the ecosphere also becomes clearer, and it is harder and harder to ignore the fact that an individual's actions have ramifications that affect a multitude of others. The interdependent-community viewpoint has lately been expressed as the Gaia hypothesis, "the theory that our planet and its creatures constitute a single self-regulating system that is in fact a great living being or organism." [42] While this view of the planet is probably too limited, dictated by the fact that humans are necessarily organism-oriented, it is certainly an improvement over a world view that recognizes no limits to the growth of human resource use. Those talking about the interconnections of the ecosphere range from poets to physicists and microbiologists, from meteorologists to ecologists and theologians. J. Stan Rowe, professor emeritus of plant ecology and crop science at the University of Saskatchewan, Saskatoon, Canada, put it this way:

> . . . as living creatures we participate in a *whole* ecological system whose integrated parts have evolved in such a close evolutionary way, over four and one half billion years, that what we call "life" is as much a property of the planet as it is of protoplasm. In no realistic sense can organisms—people, other animals, plants—be separated from the planet's enveloping matrix, which is as spirited, as animated, as they. [43]

In this time of discovering humankind's limits to growth, it is especially important to try to infuse science with an awareness of the importance of interconnections. This awareness will make a great difference in determining the kind of agriculture that will exist in the future.

For a more specific example of how values influence science and technology, consider for a moment the values today's food crops reflect. In agriculture, it is apparent that social values have influenced the genotypes of modern crop varieties. Genes for hard tomatoes, which can be mechanically harvested, tell us that mechanical expediency is more important than job security for agricultural workers or the quality of food. Crops dependent on herbicides to deal with weeds speak of the value placed on economic efficiency rather than environmental safety. Perfect winter produce in the supermarkets of the northern states is tribute to the emphasis on cosmetics, aesthetics, and comfort over humane treatment of the poor in Latin America and fair distribution of resources. A change in crop characteristics to a set more compatible with sustainability requires a simultaneous change in the values held by society.

Even the types of questions scientists seek to answer in their research are value-dictated. Most traditional agricultural and ecological research has assumed a purely competitive model and therefore expected to find trade-offs between goals. This perspective is applied even within organisms. Thus the idea of perennial grain crops has been dismissed by some as impossible because it has been assumed that a plant must trade off perenniality (allocation of energy to roots and rhizomes) to achieve high yield (allocation of energy to seed). It has been assumed that storage and reproduction are in strict competition within a plant. If, on the other hand, scientists accept the potential for cooperation and even synergism within nature, they can approach a problem by looking for patterns that work, rather than trade-offs—and they may actually find some.

As an example of how this sort of value shift would influence research, consider that researchers who have come to recognize the importance of mycorrhizal associations in plant-soil nutrient relations wish to apply this to agriculture to improve nutrient cycling. One approach, compatible with a value system honoring competition and individual gain, would be to culture individual fungi in the lab and develop marketable soil additives to enhance the amount of mycorrhizal activity in the soil.[44] This approach would likely bring fame and fortune to the scientists or companies developing the concept and the *products,* perhaps would benefit only the farmers who could afford to purchase the product, but would not reduce farmers' reliance on purchased inputs. A different approach, compatible with valuing synergism and cooperation, might be to research methods to improve mycorrhizae populations in the soil by adding appropriate sources of organic matter in the proper amounts, planting mixtures of known mycorrhizal plants, and so forth. This work would likely result in a *process* that could be used in many variations with on-farm-derived products. It would be unlikely to bring fame or fortune to any one scientist or company because it would probably have many variations and be unpatentable. It would enhance the internal recycling of nutrients and reduce reliance on outside inputs. Thus, value-based choices of research path can influence whether even an ecologically sound concept is put to use to make agriculture more sustainable.

SHIFTING VALUES TOWARD SUSTAINABILITY

Clearly, values that promote sustainability must permeate not only the research system and grassroots farmers' movements, but also the economic system and the legislative system (which sets farm policy) if sustainable agriculture is ever to become a practical reality. An agriculture is only as sustainable as the society of which it is a part. One problem here is that sustainability is not easily translated into present economic terms. Sustainability is basically about nonmonetary factors. It is about maintaining the earth's stock of resources, but also about maintaining the earth's natural *processes* so that the ecosphere can continue to function in a way to promote life, including human life. This translates to living off the fruits of the land—the production generated from the constant and predictable flow of energy from the sun—rather than living off the accumulated biological and mineral resources and capacities of the ecosphere. William Rees put this in economic terms—living off the "interest" instead of the "capital"— but perhaps this terminology is to be avoided. After all, to a great extent, the problems addressed in this book have arisen from a mindset that views agriculture as an economic rather than a biological system. Author Wendell Berry has promoted use of the term *usufruct*, the legal term for living off the interest, which is literally translated as "using the fruits."

The problem with translating sustainability into economic terms is that the present economic system is built on spending the "capital" as though it will never be used up; it is built on a continual increase in resource use. Sustainability is not about continual growth in consumption; it is about steady states. Thus, it is very difficult to adjust the economic system so that the nonmonetary factors that must become part of the accounting system in order to achieve sustainability are given an economic value. These nonmonetary factors include, on the debt side, pollution, resource depletion, and disruption of stabilizing processes. On the credit side, they include resource conservation, recycling and maintenance of resource pools, enhancement of stabilizing processes, and minimization of wastes.[45] As with any household economy, it is absolutely essential for long-term survival that humankind devise a new economic order that rewards activities that add to the credits and discourages activities that add to the debts.

It must be recognized that ultimately any economy is based on how human beings make a living here on the earth, and on all human relationships with the things of the earth—with organisms and with inert materials, with processes and cycles that link all parts of the ecosphere. Success in preserving the ecosphere may ultimately be measured in the amount of species diversity sustained. It must be recognized that human economics are a subset of nature's economy, simply because humans are part of nature. People can no longer afford the luxury of placing nature outside themselves, in the "environment." Rather, it is necessary to enter into nature's creative process. The "environment" is reaching the limits of its ability to cover people's debts by absorbing their wastes, supplying their excess energy sources beyond daily solar input, and supplying stored materials. It is even necessary to move beyond using the term *environment*, for as J. Stan Rowe put it bluntly, "'Environment' is its own putdown, its own pejorative. At face value it means what surrounds something else that is the center of interest. Environment proclaims itself to be peripheral and therefore asks not to be taken seriously."[46] An economy that includes humanity and environment as interconnected, the "Great Economy" of Wendell Berry, has as its founding principle, not unlimited production, but sustainability. As Berry put it simply, "It proposes to endure."[47] To this point, humankind has neglected to put costs or values on, or even to acknowledge and recognize, many of its charges to the Great Economy.

In an interconnected world, a little neglect goes a long way. One thing done wrong can have ramifications throughout the system. But one thing done right can solve several problems at once.[48] An obvious example is the wide-ranging consequences of pesticide use. The costs associated with these are simply not included in the price tag of the products. Yet, eventually, someone pays. When overuse of a pesticide results in resistant pests or new pests, farmers pay down the line in the form of higher pesticide costs to pay for development of new chemicals, or for more chemicals, or in the form of reduced yields. When a farm pesticide escapes into the groundwater, neighbors for miles around pay (although unknowingly, oftentimes) in ill health or in costs for new wells, water filters, or purchase of expensive bottled water. When pesticides kill stream fauna, all other species that eat fish suffer in addition to the fishing business. The true costs are diffused across human and wild communities in forms that are not easily traceable or assessable. No one in particular is held accountable for the costs; yet everyone, in general, pays in some way.

The result is a generally diminished quality of life, more tenuous relations with nonhuman nature, and less resilience in the natural system with which to right the wrongs. These are the results of externalizing the costs. In a 1979 paper, David Pimentel and colleagues tried to put monetary values on all the direct and indirect consequences of pesticide use. Then they tried to evaluate the economic consequences of not using pesticides. After many pages of figures, they concluded that the benefits were not worth the costs of using pesticides, even when these costs were in most cases grossly underestimated.[49]

Perhaps the most important nonmonetary factor that the economic system must take into account is the future. Probably the greatest failing of the money economy is the extent to which it discounts the future. It does not account for the present costs of ensuring future livelihoods. Without this factor, society is operating only on blind faith. When people do not consider whether the land will be just as healthy for their grandchildren, they are gambling with their "descendants' birthright to the natural inheritance of this planet."[50] It is all too easy to use too much now and not have enough left later on. It is clear from current farm economic problems that little value is placed on the future. This can be seen by examining the exceptions—isolated pockets of culture that clearly value the future.

In a chapter in *Farm Work and Fieldwork,* Susan Carol Rogers described one such place—Freiberg, Illinois—a community where a strong German heritage influenced farming style compared to the rest of the state. While the majority of Illinois farmers dramatically shifted from diversified farms toward a predominance of specialized cash-grain farms between 1964 and 1978, Freiberg farms maintained mixed farming, with nearly all including animals and feed grains. Also, Freiberg farms remained smaller, mostly owner-operated (70 percent compared to 45 percent statewide), and commanded higher land values. These characteristics reflect the strong cultural commitment to long-term family farming in Freiberg. Land prices stayed high because families held on to their land. One farmer commented that he wished he could come back in 500 years to see what his descendants had done with the farm. Farms became diversified livestock–feed grain operations because intensification was the only way to bring children into farming, given the lack of available land for expansion. Clearly, social values can influence the structure of farming. When a farming community has a heritage of passing down farms from generation to generation, they make decisions based on the long run, not just on short-run economic returns.

Potentially, government can take a positive role. Policies affect how much of the future is factored in, how many externalities become internalized. A simple example is water pollution legislation. Were it the government's policy to require that all users of stream water had to intake water for their own use downstream of their own effluent, costs of cleaning up the effluent would be internalized by the users. If pesticide manufacturers' license fees were scaled to reflect the real environmental costs of their products, the cost/benefit equation for pesticide use would be tipped steeply toward IPM or biological control. If farm program benefits were scaled to reward land stewardship, farmers would have more motivation and, perhaps, ability to practice good stewardship. The 1985 and 1990 farm bills took the first steps in that direction, but there is still a long way to go.

Society holds resource users accountable primarily via taxes on profits (income) and potential profits (property). Only recently have resource users begun to be held accountable for negligence and abuse. Polluters-pay laws and litigation against companies for toxic-waste dumping are examples of the latter. In Minnesota, some farmers have supported county ordinances that enforce soil conservation and hold land owners accountable for negligence of their soil when that negligence affects their neighbors' land. So far, farmers have largely escaped accountability for their application of toxins, but ongoing suburban expansion into farming country has brought this issue to the forefront, and things are beginning to change.[51] The Iowa Groundwater Protection Act discussed in Chapter 2 is a landmark example of society deciding to assign responsibility to chemical producers and users. It requires a tax on products and practices that have the potential to pollute groundwater, a tax that is meant to reflect the environmental costs of using these products and practices. The tax revenues are designated to pay for research and the changeover to a sustainable agriculture. These changes are symptomatic of a societal shift away from externalizing all nonmonetary costs and toward a return to acceptance of a responsible relationship with nature.

This discussion leads to a final point—that of humility, one last essential ingredient in the development of sustainable agriculture. This is the humility that realizes that humans are part of nature, not above it. This humility is expressed by a willingness to listen—by scientists listening to farmers, and to one another, across disciplinary bounds; by farmers listening to new ideas from scientists, as well as remaining attentive to nature and their own experience; by all of us listening humbly and carefully to history as well. This humility requires a willingness to admit to many defeats in the historical struggle with nature—defeats by insects, wind, water, and gravity. It also requires attention to the few success stories—those histories wherein farming has endured as long as human habitation, with no apparent harm done. Studying traditional agroecosystems may be the key to speeding the discovery of agroecological principles that will permit the development of sustainable agriculture throughout the world.[52]

But, most importantly, people must learn to listen to nature. After all, nature works. It is the ultimate template, the ultimate pattern that works. Humanity need not fear that as a species we will ever wipe out nature, but we can easily wipe out ourselves. Nature persists, but civilizations are ephemeral. The most essential challenge for humanity is to learn to eat from nature's bounty without destroying it in the process, to find our appropriate niche within nature. This book is not advocating a back-to-nature movement; instead, it seeks some healing ground between the technology-will-solve-everything perspective and a back-to-nature resignation. Technology has given us options that were not available to earlier generations. World population growth demands new relationships with nature. Society must put its technological prowess to work and learn to use solar energy in new ways to help accommodate the burgeoning human population. But it must recognize nature's limits and cycles and work within them to do so, and in the process may tear down the fence between human beings and the environment.

As a society we have perceived continually increasing consumption as feasible and, as a result, have placed little value on taking a long view and spent little time or energy planning for the future.

When family bonds are broken each generation by a mobile society, then the motivation to leave a legacy of soil fertility to future generations is lost, and it is not surprising that sustainability does not occupy a central theme in conventional agricultural practice. Now, as the many problems of modern agriculture surface, it is clear that these problems are multifaceted and long-term by nature, and therefore narrow, short-term solutions cannot work. But long-term solutions that are not economically feasible in the short run are not possible until society is willing to underwrite farmers' short-term losses. Similarly, solutions to problems that are not even perceived by those with narrow, short-term concerns (for example, soil erosion) will not be adopted with fervor. The challenge for the present is to learn to factor in the future, to think in the long term, and to put humans' tremendous creative power to work to design an agriculture that reflects nature's patterns and integrates itself into the natural cycles of the ecosphere, an agriculture that runs on sunshine and consumes only the fruits of the land, an agriculture that, with proper stewardship, will endure forever.

Notes

INTRODUCTION

1. Steve Murdock et al., "Demographic Characteristics of Rural Residents in Financial Distress and Social and Community Impacts of the Farm Crisis," in S. H. Murdock and F. Larry Leistritz, eds., *The Farm Financial Crisis* (Boulder: Westview Press, 1988), 115.
2. Eric Van Chantfort, "Financial Survey Shows Farmers on an Upswing Before Drought," *Farmline* (December–January 1989): 6.
3. Juliana King, "For Farm Finances: Promising Signs of a Cooling Crisis," *Farmline* (April 1987): 12.
4. F. Larry Leistritz et al., "Producer Reactions and Adaptations," in Murdock and Leistritz, *Farm Financial Crisis*, 105.
5. Ellen Banker, "How Healthy are Rural Banks?" *Farmline* (August 1986): 111.
6. Mark Brohan, "Tight Times for Rural Governments," *Farmline* (November 1986): 9.

CHAPTER 1

1. Conservation tillage, as defined by the Soil Conservation Service in 1984, was practiced on 66 million acres in 1982, but only 24 million acres when the first NRI was performed. David Schertz, "Conservation Tillage: An Analysis of Acreage Projections in the United States," *Journal of Soil and Water Conservation* 43 (1988): 257.
2. M. Gordon Wolman, "Soil Erosion and Crop Productivity: A Worldwide Perspective," in R. F. Follett and B. A. Stewart, eds., *Soil Erosion and Crop Productivity* (Madison, Wis.: ASA-CSSA-SSSA, 1985), 14.

3. David Osterman and Theresa Hicks, "Highly Erodible Land: Farmer Perceptions Versus Actual Measurements," *Journal of Soil and Water Conservation* 43 (1988): 177.

4. Osterman and Hicks, "Highly Erodible Land," 178–9.

5. R. I. Papendick et al., "Regional Effects of Soil Erosion on Crop Productivity—The Palouse Area of the Pacific Northwest," in Follett and Stewart, *Soil Erosion and Crop Productivity,* 310–11.

6. E. H. Clark, "The Off-Site Costs of Soil Erosion," *Journal of Soil and Water Conservation* 40 (1985): 19–26.

7. Sandra Batie, *Soil Erosion: Crisis in America's Croplands?* (Washington, D.C.: The Conservation Foundation, 1983), 44.

8. Wolman, "Soil Erosion," 14.

9. R. J. McCracken et al., "An Appraisal of Soil Resources in the USA," in Follett and Stewart, *Soil Erosion and Crop Productivity,* 62.

10. Edward O. Wilson, "Threats to Biodiversity," *Scientific American* 261 (1989): 111.

11. Figures from James Nations and Daniel Komer, "Rainforests and the Hamburger Society," *Environment* 25 (1983): 15. Note that "maize" is called "corn" in the United States and "maize" elsewhere. The two terms are used interchangeably, as appropriate, throughout the book.

12. D. L. Plucknett et al., "Crop Germplasm Conservation and Developing Countries," *Science* 220 (1983): 163.

13. Laurence Alderson, *The Chance to Survive: Rare Breeds in a Changing World* (London: Cameron and Taylor), 8.

14. Jan Salick and Laura C. Merrick, "Use and Maintenance of Genetic Resources: Crops and Their Wild Relatives," in C. Ronald Carroll, John H. Vandermeer and Peter M. Rosset, eds., *Agroecology* (New York: McGraw-Hill, 1990), 533.

15. Salick and Merrick, "Genetic Resources," 543.

16. David Pimentel and Wen Dazhong, "Technological Changes in Energy Use in U.S. Agricultural Production," in Carroll, Vandermeer, and Rosset, *Agroecology,* 150–55.

17. B. A. Stout et al., *Energy for World Agriculture.* FAO Agriculture Series, No. 7 (Rome: Food and Agriculture Organization of the United Nations, 1979), 47.

18. Amory Lovins et al., "Energy and Agriculture," in Wes Jackson, Wendell Berry, and Bruce Colman, eds., *Meeting the Expectations of the Land* (San Francisco: North Point Press, 1984), 70.

19. Sandra Postel, "Managing Fresh Water Supplies," in *The State of the World 1985* (Washington, D.C.: The Worldwatch Institute, 1985), 11.

20. David Pimentel et al., "Water Resources in Food and Energy," *BioScience* 32 (1982): 861–67.

21. Bruce K. Ferguson, "Wither Water? The Fragile Future of the World's Most Important Resource," *The Futurist* 17 (April 1983): 29–47.

22. Sandra Postel, *Conserving Water: The Untapped Alternative,* Worldwatch Paper 67. (Washington, D.C.: The Worldwatch Institute, 1985), 7.
23. Charles Howe and W. Ashley Ahrens, "Water Resources of the Upper Colorado River Basin: Problems and Policy Alternatives," in Mohamed T. El-Ashry and Diana C. Gibbons, eds., *Water and Arid Lands of the Western United States* (New York: Cambridge University Press, 1988), 170.
24. William Ashworth, *Nor Any Drop to Drink* (New York: Summit Books, 1982), 109.
25. Postel, *Conserving Water,* 11.
26. National Research Council, *Alternative Agriculture* (Washington, D.C.: National Academy Press, 1989), 109.
27. In metric, the figures are 14.5, 156.5, and 47 kilograms per hectare of pesticides, nitrogen, and phosphate, respectively. Figures from Michael V. McKenry, "Potential Problems with Soil Applied Chemicals," in Harold G. Alford and Mary P. Ferguson, eds., *Pesticides in Soil and Groundwater* (Berkeley: Division of Agricultural Sciences, University of California, 1983), 15–21.
28. For examples, see studies reviewed in Bijay Singh and G. S. Sekhon, "Nitrate Pollution of Groundwater from Farm Use of Nitrogen Fertilizers," *Agriculture and Environment* 4 (1978/1979): 208–9, and J. W. Moore, *Balancing the Needs of Water Use* (New York: Springer-Verlag, 1989), 105–6.
29. George Hallberg, "From Hoes to Herbicides: Agriculture and Groundwater Quality," *Journal of Soil and Water Conservation* 41 (1986): 359.
30. Singh and Sekhon, "Nitrate Pollution," 214.
31. Tom Addiscott, "Farmers, Fertilisers and the Nitrate Flood," *New Scientist* 120 (1988): 54.
32. Larry Fruhling, "Please Don't Drink the Water," The Progressive 50 (1986): 32–33.
33. Singh and Sekhon, "Nitrate Pollution," 216–17.
34. Hallberg, "From Hoes to Herbicides," 359.
35. Mahfouz Zaki et al., "Pesticides in Groundwater: The Aldicarb Story in Suffolk County, NY," *American Journal of Public Health* 72 (1982): 1393–94.
36. "The Persistence of EDB in Soils," *Environment* 29 (no. 10, 1987): 24.
37. James W. Moore, *Balancing the Needs of Water Use,* 119.
38. C. Shannon Stokes and Kathy Brace, "Agricultural Chemical Use and Cancer Mortality in Selected Rural Counties in the U.S.A.," *Journal of Rural Studies* 4 (1988): 245.
39. Leon Burmeister et al., "Selected Cancer Mortality and Farm Practices in Iowa," *American Journal of Epidemiology* 118 (1983): 72.
40. Kenneth Cantor and Aaron Blair, "Farming and Mortality from Multiple Myeloma: A Case-Control Study with the Use of Death Certificates," *Journal of the National Cancer Institute* 782 (1984): 252.
41. Moore, *Balancing the Needs of Water Use,* 121.

42. Wendy Wall, "Iowa Bill to Curb Groundwater Pollution Could Prompt Similar Steps Elsewhere," *Wall Street Journal*, 13 March 1987, 18.
43. Molly Joel Coye, "The Health Effects of Agricultural Production: I. The Health of Agricultural Workers," *Journal of Public Health Policy* 6 (1985): 359.
44. Coye, "The Health of Agricultural Workers," 363.
45. Angus Wright, "Rethinking the Circle of Poison: The Politics of Pesticide Poisoning among Mexican Farm Workers," *Latin American Perspectives* 13 (1986): 35.
46. Marjorie Sun, "EDB Contamination Kindles Federal Action," *Science* 223 (1984): 464.
47. "Pillsbury Wins Insurance Ruling," *New York Times*, 9 December 1988, p. D4, col. 1.
48. See "EDB's Long-Lasting Legacy," *Science News* 127 (1985): 313.
49. Dixie Farley, "Setting Safe Limits on Pesticide Residues," *FDA Consumer* 22 (1988): 11.
50. Louise Fenner, "A Hard Look at What We're Eating," *FDA Consumer* 18 (1984): 8–13.
51. Figures on India are from V. P. Gupta, "Pesticide Misuse in India," *The Ecologist* 16 (1986): 36–39. Figures from the United States are from Robert Murphy and Clair Harvey, "Residues and Metabolites of Selected Persistent Halogenated Hydrocarbons in Blood Specimens from a General Population Survey," *Environmental Health Perspectives* 60 (1985): 115–20. Examples of marketplace surveys showing excessive pesticide residues from Kenya to Sri Lanka are found in David Bull, *A Growing Problem: Pesticides and the Third World Poor* (Oxford: Oxfam, 1982).
52. Gupta, "Pesticide Misuse in India," 36.
53. These figures were published in 1986 by George Georghiou in "The Magnitude of the Resistance Problem," in National Resource Council, *Pesticide Persistence: Strategies and Tactics for Management* (Washington, D.C.: National Academy Press, 1986), 18. In a letter to *Science*, published April 19, 1991, David Pimentel reported these dramatically higher figures: 504 species of insects and mites, 273 species of weeds, and 150 species of pathogens had been reported resistant to pesticides.
54. David Pimentel et al., "An Environmental Risk Assessment of Biological and Cultural Controls for Organic Agriculture," in W. Lockeretz, ed., *Environmentally Sound Agriculture* (New York: Praeger, 1983), 73.
55. The examples in the following three paragraphs were gleaned from Gordon R. Conway's *Pesticide Resistance and World Food Production* (London: Imperial College Centre for Environmental Technology, 1982) and the FAO *Report on Pest Resistance to Pesticides and Crop Loss Assessment* (Rome: Food and Agriculture Organization of the United Nations, 1979). See also Georghiou, "The Magnitude of the Resistance Problem," for other examples of cross-resistance in pests.

56. For many examples of case histories of these phenomena, see David Pimentel, *Ecological Effects of Pesticides on Non-target Species* (Washington, D.C.: U.S. Government Printing Office, 1971) and Paul DeBach, *Biological Control by Natural Enemies* (London: Cambridge University Press, 1974).

57. T. H. Coaker, "Crop Pest Problems Resulting from Chemical Control," in J. M. Cherrett and G. R. Sagar, eds., *Origins of Pest, Parasite, Disease and Weed Problems* (Oxford: Blackwell Scientific Publications, 1977), 314.

58. Francis Chaboussou, though not widely known in the U.S. literature, has published widely on this subject in Europe. For example, see "How Pesticides Increase Pests," *The Ecologist* 16 (1986): 29–35.

59. Coaker, "Crop Pest Problems," 323.

CHAPTER 2

1. Tillage erosion is the movement of soil downhill by the process of tillage. See James Shepherd, "Soil Conservation in the Pacific Northwest Wheat-producing Areas: Conservation in a Hilly Terrain," in Douglas Helms and Susan L. Flader, eds., *Agricultural History* (Berkeley: University of California Press, 1985), 139.

2. Fred P. Miller, Wayne D. Rasmussen, and L. Donald Meyer, "Historical Perspective of Soil Erosion in the United States," in R. F. Follett and B. A. Stewart, eds., *Soil Erosion and Crop Productivity* (Madison, Wis.: ASA-CSSA-SSSA, 1985), 26.

3. These percentages were expected to decline to 40 percent for worldwide energy and 48 percent for North America by 1985. Figures from B. A. Stout et al., *Energy for World Agriculture*, FAO Agriculture Series, No. 7 (Rome: Food and Agriculture Organization of the United Nations, 1979), 49.

4. Amory Lovins, Hunter Lovins, and Marty Bender, "Energy and Agriculture," in Wes Jackson, Wendell Berry, and Bruce Colman, eds., *Meeting the Expectations of the Land* (San Francisco: North Point Press, 1984), 72.

5. Eric Van Chantfort, "Agriculture and the Larger Economy: When Links Become Liabilities," *Farmline* (July 1984): 14–17.

6. John Patrick Jordan, Paul F. O'Connell, and Roland R. Robinson, "Historical Evolution of the State Agricultural Experiment Station System," in Kenneth A. Dahlberg, ed., *New Directions for Agriculture and Agricultural Research* (Totowa, N.J.: Rowman and Allanheld, 1986), 147–62.

7. For example, see L. W. Kannenberg, "Utilization of Genetic Diversity in Crop Breeding," in C. W. Yeatman, D. Kafton, and G. Wilkes, eds., *Plant Genetic Resources: A Conservation Imperative* (Boulder: Westview Press, 1984), 93–109.

8. Richard Lewontin and Jean-Pierre Barlan, "The Political Economy of Ag-

ricultural Research: The Case of Hybrid Corn," in C. Ronald Carroll, John H. Vandermeer, and Peter M. Rosset, eds., *Agroecology* (New York: McGraw-Hill, 1990), 613–28.

9. A further discussion of these premises and their current validity can be found in the section "Agribusiness and Research" below and in Chapter 6.

10. Paul O'Connell, "National Agenda Setting and the State Agricultural Experiment Stations," in the Experiment Station Committee on Organization and Policy, *Research Perspective* (Cooperative State Research Service, 1985), 44. Note that this view is not accepted by all. In fact, food is generally a rather inelastic market because demand is related to population size and the general state of the economy.

11. Richard Levins and Richard Lewontin, *The Dialectical Biologist* (Cambridge: Harvard University Press, 1985), 211.

12. Don F. Hadwiger, *The Politics of Agricultural Research* (Lincoln: University of Nebraska Press, 1982), 97. This pattern is changing, with universities now trying to reap the patents and profits of biotechnology research. See Chapter 6 for a more detailed discussion.

13. Hadwiger, *Politics of Agricultural Research*, 99.

14. Lawrence Busch and William Lacy, "Agricultural Research in Developed Nations," in Carroll, Vandermeer, and Rosset, *Agroecology*, 570.

15. Lawrence Busch and William Lacy, *Science, Agriculture, and the Politics of Research* (Boulder: Westview Press, 1983), 164.

16. This paragraph draws on the description of this research found on pages 216–20 in Marty Strange's *Family Farming: A New Economic Vision* (Lincoln: University of Nebraska Press, 1988).

17. Strange, *Family Farming*, 217.

18. Strange, *Family Farming*, 220.

19. Recently, many critiques of agricultural research institutions have detailed these sources of inertia. See, for example, the extensive research reported in Busch and Lacy's *Science, Agriculture, and the Politics of Research*.

20. Stanislaus J. Dundon, "Hidden Obstacles to Creativity in Agricultural Science," *Twentieth Century Agricultural Science*, n.s., vol. 8 (1983): 109.

21. The following description is taken from Dundon, "Hidden Obstacles to Creativity," 112–13.

22. Busch and Lacy, "Agricultural Research in Developed Nations," 568.

23. Rod MacRae et al., "Agricultural Science and Sustainable Agriculture: A Review of the Existing Scientific Barriers to Sustainable Food Production and Potential Solutions," *Biological Agriculture and Horticulture* 6 (1989): 182.

24. Evelyn Fox Keller, *Reflections on Gender and Science* (New Haven: Yale University Press, 1985), 36.

25. Patrick Madden, "Beyond Conventional Economics—An Examination of the Values Implicit in the Neoclassical Economic Paradigm as Applied to the Evaluation of Agricultural Research," in Dahlberg, *New Directions for Agriculture and Agricultural Research.*

26. Miller, Rasmussen, and Meyer, "Historical Perspective on Soil Erosion," 26.

27. Shiraz Vira and Harry Riehle, "Conservation in the Palouse: An Economic Dilemma," in W. E. Jeske, ed., *Economics, Ethics, Ecology: Roots of Productive Conservation* (Ankeny, Iowa: Soil Conservation Society of America, 1981), 453.

28. MacRae et al., "Agricultural Science and Sustainable Agriculture," 177.

29. For example, see studies in the bibliography by John L. Harper and Ola Inghe with Carl Olaf Tamm.

30. David R. Foster, "Disturbance History, Community Organization and Vegetation Dynamics of the Old-growth Pisgah Forest, Southwestern New Hampshire, U.S.A.," *Journal of Ecology* 76 (1988): 105–34.

31. John E. Weaver, *Prairie Plants and Their Environment* (Lincoln: University of Nebraska Press, 1968).

CHAPTER 3

1. Howard T. Odum is a leading proponent of this view. See, for example, his paper "Self-organization, Transformity, and Information," *Science* 242 (1988): 1132–39.

2. James A. Teeri and L. G. Stowe, "Climatic Patterns and the Distribution of C_4 Grasses in North America," *Oecologia* 23 (1976): 1–12.

3. Nona Chiarello, James C. Hickman, and Harold A. Mooney, "Endomycorrhizal Role for Interspecific Transfer of Phosphorus in a Community of Annual Plants," *Science* 217 (1982): 941–43; Christopher van Kessel, Paul W. Singleton, and Heinz J. Hoben, "Enhanced N-transfer from Soybean to Maize by Vesicular Arbuscular Mycorrhizal (VAM) Fungi," *Plant Physiology* 79 (1985): 562–63; D. M. Eissenstat, "A Comparison of Phosphorus and Nitrogen Transfer Between Plants of Different Phosphorus Status," *Oecologia* 82 (1990): 342–47.

4. Peter W. Price provides an excellent review of soil ecosystem dynamics in an evolutionary context in "An Overview of Organismal Interactions in Ecosystems in Evolutionary and Ecological Time," *Agriculture, Ecosystems and Environment* 24 (1988): 369–77.

5. The concept of the climax community as a constant, self-perpetuating, final phase of succession was proposed by Frederick E. Clements in "Nature and Structure of the Climax," *Journal of Ecology* 24 (1936): 252–84, and developed by Eugene P. Odum in "The Strategy of Ecosystem Development," *Science* 164 (1969): 262–70.

6. Although the exact mechanisms of succession are controversial, the general pattern is usually predictable. For further exploration of some ideas, see Joseph H. Connell and Ralph O. Slatyer, "Mechanisms of Succession in Natural Communities and Their Role in Community Stability and Organization," *American Naturalist* 111 (1977): 1119–44; David Tilman, *Plant Strategies and the Dynamics and Structure of Plant Communities*, (Princeton: Princeton University Press, 1988); Daniel B. Botkin, *Discordant Harmonies* (New York: Oxford University Press, 1990).

7. Definition from C. S. Holling, "Resilience and Stability of Ecological Systems," *Annual Review of Ecology and Systematics* 4 (1973): 1–23.

8. Many of the concepts presented in this section build upon the ideas of David J. Rapport in "What Constitutes Ecosystem Health?" *Perspectives in Biology and Medicine* 33 (1989): 120–32.

9. F. Herbert Bormann, "Air Pollution and Forests: An Ecosystem Perspective," *BioScience* 35 (1985): 434–41.

10. See, for example, Bill Freedman and T. C. Hutchinson, "Long Term Effects of Smelter Pollution at Sudbury, Ontario, on Surrounding Forest Communities," *Canadian Journal of Botany* 58 (1980): 2123–40.

11. Jack R. Harlan, "Diseases as a Factor in Plant Evolution," *Annual Review of Phytopathology* 14 (1976): 31–51; A. Dinoor and N. Eshed, "The Role and Importance of Pathogens in Natural Plant Communities," *Annual Review of Phytopathology* 22 (1984): 443–66.

12. Of course, and as Daniel Botkin points out in *Discordant Harmonies*, both population growth rate and carrying capacity of the environment typically fluctuate in nature. Thus, for resource managers to determine exactly a maximum sustainable yield for a wild population is difficult, if not impossible.

13. For an example, see O. Jennersten, S. G. Nilsson, and U. Wastljung, "Local Plant Populations as Ecological Islands: The Infection of *Viscaria vulgaris* by the Fungus *Ustilago violacea*," *Oikos* 41 (1983): 391–95.

14. Robert H. Whittaker and Simon A. Levin, "The Role of Mosaic Phenomena in Natural Communities," *Theoretical Population Biology* 12 (1977): 117–39.

15. Examples where periodic disturbance maintains higher species diversity exist for marine, forest, and grassland ecosystems. See Joseph H. Connell, "Diversity in Tropical Rain Forests and Coral Reefs," *Science* 199 (1978): 1302–10; Scott L. Collins, "Interaction of Disturbances in Tallgrass Prairie: A Field Experiment," *Ecology* 68 (1987): 1243–50.

16. Although this idea is by no means universally held among ecologists, support for it exists in the following papers: James H. Connell, "Diversity and the Coevolution of Competitors, or the Ghost of Competition Past," *Oikos* 35 (1980): 131–38; Judith A. D. Parrish and Fakhri A. Bazzaz, "Competitive Interactions in Plant Communities of Different Successional Ages," *Ecology* 63 (1982): 314–20.

17. These theories were first presented in Paul Feeny, "Plant Apparency and Chemical Defense," *Recent Advances in Phytochemistry* 10 (1976); 1–39; David F. Rhoades and Rex G. Cates, "Toward a General Theory of Plant Antiherbivore Chemistry," *Recent Advances in Phytochemistry* 10 (1976): 168–213.
18. *Sensu* R. Dawkins and J. Krebs, "Arms Races Between and Within Species," *Proceedings of the Royal Society of London*, ser. B, 205 (1979): 489–511.
19. Examples from natural populations are found in F. W. Cobb, Jr., et al., "Rate of Spread of *Ceratocystis wagneri* in Ponderosa Pine Stands in the Central Sierra Nevada," *Phytopathology* 72 (1982): 1359–62; Carol K. Augspurger, "Seed Dispersal of the Tropical Tree, *Platypodium elegans*, and the Escape of its Seedlings from Fungal Pathogens," *Journal of Ecology* 71 (1983): 759–71; Carol K. Augspurger, "Seedling Survival of Tropical Tree Seedlings: Interactions of Dispersal Distance, Light-gaps and Pathogens," *Ecology* 65 (1984): 1705–12; Helen M. Alexander, "Spatial Patterns of Disease Introduced by *Fusarium moniliforme* var *subglutinans* in a Population of *Plantago lanceolata*," *Oecologia* 62 (1984): 141–43. Examples from agricultural plots are found in Jeremy J. Burdon, *Diseases and Plant Population Biology* (Cambridge: Cambridge University Press, 1987); Jeremy J. Burdon and G. A. Chilvers, "Host Density as a Factor in Plant Disease Ecology," *Annual Review of Phytopathology* 20 (1982): 143–66.
20. C. C. Mundt and K. J. Leonard, "Effect of Host Genotype Unit Area on Epidemic Development of Crown Rust Following Focal and General Inoculations of Mixtures of Immune and Susceptible Host Plants," *Phytopathology* 75 (1985): 1141–45; C. C. Mundt and K. J. Leonard, "Effect of Host Genotype Unit Area on Development of Focal Epidemics of Bean Rust and Common Maize Rust in Mixtures of Resistant and Susceptible Plants," *Phytopathology* 76 (1986): 895–900.
21. C. Forsyth and N. A. C. Brown, "Germination of the Dimorphic Fruits of *Bidens pilosa* L." *New Phytologist* 90 (1982): 151–64; Jon K. Piper, "Germination and Growth of Bird-dispersed Plants: Effects of Seed Size and Light on Seedling Vigor and Biomass Allocation," *American Journal of Botany* 73 (1986): 959–65.
22. Paul Dann, "Toward Self-sufficient Food Production," *Agricultural Science* (June–July 1988): 16–19. For statistics on the amount of energy expended per ton of grain produced in the United States, see Lester Brown, "The Vulnerability of Oil-based Farming," *World Watch* 1(2) (1988): 24–29.
23. David Pimentel, "Energy Flow in Agroecosystems," in Richard Lowrance, Benjamin R. Stinner, and Garfield J. House, eds., *Agricultural Ecosystems* (New York: John Wiley and Sons, 1984), 121–32.
24. Clive A. Edwards and J. F. Lofty, "The Influence of Cultivation on Animal Populations," in J. Vanek, ed., *Progress in Soil Biology* (Prague: Academia Publications, 1975), 399–408.

25. H. J. Haas, C. E. Evans, and E. F. Miles, "Nitrogen and Carbon Changes in Great Plains Soils as Influenced by Cropping and Soil Treatments," USDA Technical Bulletin 1164 (Washington, D.C.: 1957); D. W. Anderson and David C. Coleman, "The Dynamics of Organic Matter in Grassland Soils," *Journal of Soil and Water Conservation* 40 (1985): 211–16.

26. L. E. Woods, "Active Organic Matter Distribution in the Surface 15 Cm of Undisturbed and Cultivated Soil," *Biology and Fertility of Soils* 8 (1989): 271–78.

27. J. A. Hobbs and P. L. Brown, *Nitrogen and Organic Carbon Changes in Cultivated Western Kansas Soils*, Kansas Agricultural Experiment Station Bulletin 89 (Manhattan, Kans.: Kansas State University, 1957).

28. W. E. Larson, F. J. Pierce, and R. H. Dowdy, "The Threat of Soil Erosion to Long-term Crop Production," *Science* 219 (1983): 458–65.

29. See, for example, Jeremy J. Burdon and D. R. Marshall, "Inter- and Intra-specific Diversity in the Disease-response of *Glycine* Species to the Leaf Rust Fungus *Phakopsora pachyrhizi*," *Journal of Ecology* 69 (1981): 381–90; Jeremy J. Burdon, J. D. Oates, and D. R. Marshall, "Interactions Between *Avena* and *Puccinia* species. I. The Wild Hosts: *Avena barbata* Pott ex Link, *A. Fatus* L., *A. ludoviciana* Durieu," *Journal of Applied Ecology* 20 (1983): 571–84; A. Segal, J. Manisterski, G. Fishbeck, and I. Wahl, "How Plant Populations Defend Themselves in Natural Ecosystems," in James G. Horsfall and Ellis B. Cowling, eds., *Plant Disease: An Advanced Treatise. Vol. 5. How Plants Defend Themselves.* (New York: Academic Press, 1980), 75–102; D. E. Zimmer and D. Rehder, "Rust Resistance of Wild *Helianthus* Species of the North Central United States," *Phytopathology* 66 (1976): 208–11.

CHAPTER 4

1. See, for example, William Cronon, *Changes in the Land* (New York: Hill & Wang, 1983).

2. Excellent recent reviews of integrated pest management are Miguel A. Altieri, "Agroecology: A New Research and Development Paradigm for World Agriculture," *Agriculture, Ecosystems and Environment* 27 (1989): 37–46; David A. Andow and Peter M. Rosset, "Integrated Pest Management," in C. Ronald Carroll, John H. Vandermeer, and Peter M. Rosset, eds., *Agroecology* (New York: McGraw-Hill, 1990), 413–39.

3. Robert F. Luck, B. M. Shepard, and P. E. Kenmore, "Experimental Methods for Evaluating Arthropod Natural Enemies," *Annual Review of Entomology* 33 (1988): 367–91.

4. See, for example, P. J. Goodman and M. Collison, "Varietal Differences in Uptake of ^{32}P Labelled Phosphate in Clover Plus Ryegrass Swards and Monocultures," *Annals of Applied Biology* 100 (1982): 559–65.

5. Excellent reviews of intercropping can be found in R. I. Papendick,

P. A. Sanchez, and G. B. Triplett, eds., *Multiple Cropping*, ASA Special Publication No. 27 (Madison, Wis.: American Society of Agronomy, Crop Science Society of America, and Soil Science Society of America, 1976); Charles A. Francis, ed., *Multiple Cropping Systems* (New York: Macmillan, 1986); John H. Vandermeer, *The Ecology of Intercropping* (Cambridge: Cambridge University Press, 1989).

6. Richard B. Root, "Organization of a Plant-arthropod Association in Simple and Diverse Habitats: The Fauna of Collards (*Brassica oleracea*)," *Ecological Monographs* 43 (1973): 95–124; Stephen J. Risch, David Andow, and Miguel A. Altieri, "Agroecosystem Diversity and Pest Control: Data, Tentative Conclusions, and New Research Directions," *Environmental Entomology* 12 (1983): 625–29.

7. J. G. Smith, "Influence of Crop Background on Natural Enemies of Aphids on Brussels Sprouts," *Annals of Applied Biology* 83 (1976): 15–29; D. J. Horn, "Effect of Weedy Backgrounds on Colonization of Collards by Green Peach Aphid, *Myzus persicae*, and Its Major Predators," *Environmental Entomology* 10 (1981): 285–89; G. E. Brust, Benjamin R. Stinner and D. A. McCartney, "Predator Activity and Predation in Corn Agroecosystems," *Environmental Entomology* 15 (1986): 1017–21.

8. T. K. Kroll, L. D. Moore, and G. H. Lacy, "Interplanting Susceptible and Resistant Radish Cultivars Reduces Colonization by *Plasmodiophora brassicae*" *HortScience* 19 (1984): 403–4; Jeremy J. Burdon, *Diseases and Plant Population Biology* (Cambridge: Cambridge University Press, 1987).

9. Elroy L. Rice, *Allelopathy*, 2nd ed. (Orlando: Academic Press, 1984).

10. Matthew Liebman, "Ecological Suppression of Weeds in Intercropping Systems: A review," in Miguel A. Altieri and Matthew Liebman, eds., *Weed Management in Agroecosystems: Ecological Approaches* (Boca Raton, Fla.: CRC Press, 1988), 198–212.

11. Brust, Stinner, and McCartney, "Predator Activity and Predation in Corn Agroecosytems," 1017–21.

12. Garfield R. House and Benjamin R. Stinner, "Agriculture and Ecosystems," in Patricia Allen and Debra Van Dusen, eds., *Global Perspectives on Agroecology and Sustainable Agricultural Systems* (Santa Cruz: Agroecology Program, University of California, 1988), 345–54.

13. R. F. Follett and G. A. Peterson, "Surface Soil Nutrient Distribution as Affected by Wheat-fallow Tillage Systems," *Soil Science Society of America Journal* 52 (1988): 141–47.

14. Amory Lovins, L. Hunter Lovins, and Marty Bender, "Energy and Agriculture," in Wes Jackson, Wendell Berry, and Bruce Colman, eds., *Meeting the Expectations of the Land*, (San Francisco: North Point Press, 1984), 68–86.

15. J. L. Baker and J. M. Laflen, "Water Quality Consequences of Conservation Tillage," *Journal of Soil and Water Conservation* 38 (1983): 186–93;

Maureen K. Hinkle, "Problems with Conservation Tillage," *Journal of Soil and Water Conservation* 38 (1983): 201–6.

16. Miguel A. Altieri, "Why Study Traditional Agriculture?" in Carroll, Vandermeer, and Rosset, *Agroecology*, 551–64.

17. Our discussion of desert soils is drawn largely from James A. MacMahon, "Warm Deserts," in Michael G. Barbour and William Dwight Billings, eds., *North American Terrestrial Vegetation* (Cambridge: Cambridge University Press, 1988), 231–64, and references therein.

18. John N. Thompson and Richard N. Mack, "Evolution in Steppe with Few Large, Hooved Mammals," *American Naturalist* 119 (1982): 757–73.

19. Gary Paul Nabhan, *The Desert Smells Like Rain* (San Francisco: North Point Press, 1982); Gary Paul Nabhan, *Gathering the Desert* (Tucson: University of Arizona Press, 1985).

20. W. F. Harris et al., "Analysis of Carbon Flow and Productivity in a Temperate Deciduous Forest Ecosystem," in *Productivity of World Ecosystems* (Washington, D.C.: National Academy of Sciences, 1975), 116–22.

21. James Reade Runkle, "Patterns of Disturbance in Some Old-growth Mesic Forests in Eastern North America," *Ecology* 63 (1982): 1533–46.

22. Cronon, *Changes in the Land*.

23. J. Russell Smith, *Tree Crops: A Permanent Agriculture* (New York: The Devin-Adair Co., 1953).

24. Alabama Agricultural Experiment Station, *Fifty-third Annual Report of the Alabama Agricultural Experiment Station of the Alabama Polytechnic Institute* (Auburn, Ala.: Alabama Polytechnic Institute, 1942); Smith, *Tree Crops*.

25. Gary S. Hartshorn, "Neotropical Forest Dynamics," *Tropical Succession (Biotropica* supplement) 12 (1980): 23–30.

26. John Ewel et al., "Slash and Burn Impacts on a Costa Rican Wet Forest Site," *Ecology* 62 (1981): 816–29.

27. Cory W. Berish and John J. Ewel, "Root Development in Simple and Complex Tropical Successional Ecosystems," *Plant and Soil* 106 (1988): 73–84.

28. John J. Ewel, Maria J. Mazzarino, and Cory W. Berish in personal communication with authors.

29. John Beer, "Litter Production and Nutrient Cycling in Coffee (*Coffea Arabica*) or Cacao (*Theobroma Cacao*) Plantations with Shade Trees," *Agroforestry Systems* 7 (1988): 103–14.

30. Barbara A. Daniels Hetrick, D. Gerschefske Kitt, and G. Thompson Wilson, "Mycorrhizal Dependence and Growth Habit of Warm-season and Cool-season Tallgrass Prairie Plants," *Canadian Journal of Botany* 66 (1988): 1376–80; Barbara A. Daniels Hetrick, G. Thompson Wilson, and David C. Hartnett, "Relationship Between Mycorrhizal Dependence and Competitive Ability of Two Tallgrass Prairie Grasses," *Canadian Journal of Botany* 67 (1989): 2608–15.

31. Timothy R. Seastedt, Rosemary A. Ramundo, and David C. Hayes, "Maximization of Densities of Soil Animals by Foliage Herbivory: Empirical Evidence, Graphical and Conceptual Models," *Oikos* 51 (1988): 243–48.

32. K. Paustian et al., "Carbon and Nitrogen Budgets of Four Agro-ecosystems with Annual and Perennial Crops, With and Without N Fertilization," *Journal of Applied Ecology* 27 (1990): 60–84.

33. J. M. Maass, C. F. Jordan, and J. Sarukhan, "Soil Erosion and Nutrient Losses in Seasonal Tropical Agroecoystems under Various Management Techniques," *Journal of Applied Ecology* 25 (1988): 595–607.

34. The Land Institute's permanent research staff consists of Peter A. Kulakow, plant breeder; Mary K. Handley, plant pathologist; and Jon K. Piper, ecologist.

35. Charles A. Francis, C. A. Flor, and S. R. Temple, "Adapting Varieties for Intercropping Systems in the Tropics," in Papendick, Sanchez, Triplett, *Multiple Cropping*, 235–53.

CHAPTER 5

1. These four questions were first posed by Wes Jackson, *New Roots for Agriculture*, new ed. (Lincoln: University of Nebraska Press, 1985).

2. Donald L. Plucknett and Nigel J. H. Smith, "Historical Perspectives on Multiple Cropping," in Charles A. Francis, ed., *Multiple Cropping Systems* (New York: Macmillan, 1986), 20–39.

3. V. L. Bohrer, "The Prehistoric and Historic Role of the Cool-season Grasses in the Southwest," *Economic Botany* 29 (1975): 199–207; J. F. Doebley, "'Seeds' of Wild Grasses: A Major Food of Southwestern Indians," *Economic Botany* 38 (1984): 52–64; Kelly Kindscher, *Edible Wild Plants of the Prairie* (Lawrence: University Press of Kansas, 1987); S. B. Yensen and C. W. Weber, "Protein Quality of *Distichlis palmeri* Grain, a Saltgrass," *Nutrition Reports International* 35 (1987): 963–72.

4. Jack R. Harlan, "Wild Grass Seeds as Food Sources in the Sahara and Sub-Sahara," *Sahara* (2/1989): 69–74.

5. R. Reimann-Philipp, "Perennial Spring Rye as a Crop Alternative," *Journal of Agronomy and Crop Science* 157 (1986): 281–85.

6. Yensen and Weber, "Protein Quality of *Distichlis palmeri*."

7. J. Russell Smith, *Tree Crops: A Permanent Agriculture* (New York: The Devin-Adair Co., 1953).

8. A. Benzioni and R. L. Dunstone, "Jojoba: Adaptation to Environmental Stress and Implications for Domestication," *Quarterly Review of Biology* 61 (1986): 177–99; D. Pasternak et al., "Development of New Arid Zone Crops for the Negev Desert of Israel," *Journal of Arid Environments* 11 (1986): 37–59.

9. Peggy Wagoner, *Perennial Grain Research at the Rodale Research Center, 1987 Summary* (Emmaus, Pa.: Rodale Press, 1988).

10. A. I. Derzhavin, "The Theory and Practice of Producing Perennial Rye Varieties," in N. V. Tsitsin, ed., *Wide Hybridization in Plants* (Moscow: Academy of Sciences of the U.S.S.R. All-Union Academy of Agricultural Sciences, 1960), 143–52. (Publiished for the National Science Foundation, Washington, D.C., by the Israeli Program for Scientific Translations, Jerusalem, 1962.)

11. Jack R. Harlan and R. M. Ahring, *Caddo Switchgrass*, Oklahoma Agricultural Experiment Station Bulletin B-516 (Stillwater, Okla.: Oklahoma State University, 1958).

12. Robert M. Ahring, *The Management of Buffalograss Buchloe dactyloides (Nutt.) Engelm. for Seed Production in Oklahoma*, Oklahoma Agricultural Experiment Station Technical Bulletin T-109 (Stillwater, Okla.: Oklahoma State University, 1964).

13. H. R. Brown, "Growth and Seed Yields of Native Prairie Plants in Various Habitats of the Mixed Prairie," *Transactions of the Kansas Academy of Science* 46 (1943): 87–99; H. W. Cooper, "Producing and Harvesting Grass Seed in the Great Plains," *USDA Farmer's Bulletin No. 2112* (Washington, D.C.: U.S. Government Printing Office, 1957); Ahring, "The Management of Buffalograss."

14. William A. Wheeler, *Grassland Seeds* (New York: D. van Nostrand, 1957); A. Thornberg, *Grass and Legume Seed Production in Montana*, Montana State University Agricultural Experiment Station Bulletin 333 (Bozeman: Montana State University, 1971).

15. Kent V. Flannery, "The Origins of Agriculture," *Annual Review of Anthropology* 2 (1973): 271–310.

16. Jack R. Harlan and Daniel Zohary, "Distribution of Wild Wheats and Barley," *Science* 153 (1966): 1074–80.

17. Louis F. Pitelka, Sandra B. Hansen, and Jeffery W. Ashmun, "Population Biology of *Clintonia borealis*. I. Ramet and Patch Dynamics," *Journal of Ecology* 73 (1985): 169–83; E. G. Reekie anhd Fakhri A. Bazzaz, "Reproductive Effort in Plants. 3. Effect of Reproduction on Vegetative Activity," *American Naturalist* 129 (1987): 907–19; Carol C. Horvitz and Douglas W. Schemske, "Demographic Cost of Reproduction in a Neotropical Herb: An Experimental Field Study," *Ecology* 69 (1988): 1741–45.

18. Of course, not all crop mixtures are compatible and thus overyield. The purpose here is to demonstrate that many carefully designed mixtures of crops can overyield considerably.

19. Many of the overyielding examples in this section are from Brian R. Trenbath, "Plant Interactions in Mixed Crop Communities," in R. I. Papendick, P. A. Sanchez, and G. B. Triplett, eds., *Multiple Cropping*, ASA Special Publication No. 27 (Madison, Wis.: ASA-CSSA-SSSA 1976), 129–69.

20. P. Christie, E. I. Newman, and R. Campbell, "Grassland Species Can Influence the Abundance of Microbes on Each Other's Roots," *Nature* 250 (1974): 570–71.
21. P. N. Drolsom and D. Smith, "Adapting Species for Forage Mixtures," in Papendick, Sanchez, and Triplett, *Multiple Cropping*, 223–34.
22. Frank L. Barnett and Gary L. Posler, "Performance of Cool-season Perennial Grasses in Pure Stand and in Mixtures with Legumes," *Agronomy Journal* 75 (1983): 582–86; Laura S. Brophy, G. H. Heichel, and M. P. Russelle, "Nitrogen Transfer from Forage Legumes to Grass in a Systematic Planting Design," *Crop Science* 27 (1987): 753–58; A. F. Hunter and L. W. Aarssen, "Plants Helping Plants," *BioScience* 38 (1988): 34–40; Robert Becker et al., "Compositional, Nutritional, and Functional Evaluation of Intermediate Wheatgrass (*Thinopyrum intermedium*)," *Journal of Food Processing and Preservation* 15 (1991): 63–77.
23. R. S. Nauta, "Agricultural Production in the Netherlands by Natural Supply of Nitrogen," *Biological Agriculture and Horticulture* 4 (1987): 181–201.
24. Matthew Liebman, "Ecological Suppression of Weeds in Intercropping Systems: A Review," in Miguel A. Altieri and Matthew Liebman, eds., *Weed Management in Agroecosystems: Ecological Approaches* (Boca Raton, Fla.: CRC Press, 1988), 198–212.
25. See examples in Plucknett and Smith, "Historical Perspectives."
26. H. F. van Emden and G. F. Williams, "Insect Stability and Diversity of Agroecosystems," *Annual Review of Entomology* 19 (1974): 455–75.
27. Stephen. J. Risch, David Andow, and Miguel A. Altieri, "Agroecosystem Diversity and Pest Management: Data, Tentative Conclusions and New Research Directions," *Environmental Entomology* 12 (1983): 625–29.
28. E. M. Tukahira and T. H. Coaker, "Effect of Mixed Cropping on Some Insect Pests of Brassicas; Reduced *Brevicoryne brassicae* Infestations and Influences on Epigeal Predators and the Disturbance of Oviposition Behavior in *Delia brassicae*," *Entomologia Experimentalis et Applicata* 32 (1982): 129–40.
29. M. W. Adams, A. H. Ellingbae, and E. C. Rossineau, "Biological Uniformity and Disease Epidemics," *BioScience* 21 (1971): 1067–70; J. G. Horsfall et al., *Genetic Vulnerability of Major Crops* (Washington, D.C.: National Academy of Sciences, 1972); J. A. Barrett, "The Evolutionary Consequences of Monoculture," in J. A. Bishop and L. M. Cook, eds., *Genetic Consequences of Man-made Change* (London: Academic Press, 1981), 209–48.
30. T. K. Kroll, L. D. Moore, and G. H. Lacy, "Interplanting Susceptible and Resistant Radish Cultivars Reduces Colonization by *Plasmodiophora brassicae*," *HortScience* 19 (1984): 403–4.
31. Miguel A. Altieri and Matthew Liebman, "Insect, Weed, and Plant Disease Management in Multiple Cropping Systems," in Charles A. Fran-

cis, ed., *Multiple Cropping Systems* (New York: Macmillan, 1986), 183–218; K. J. Leonard, "Factors Affecting Rates of Stem Rust in Increase in Mixed Plantings of Susceptible and Resistant Oat Varieties," *Phytopathology* 59 (1969): 1845–50; Jeremy J. Burdon and R. Whitbread, "Rates of Increase of Barley Mildew in Mixed Stands of Wheat and Barley," *Journal of Applied Ecology* 16 (1979): 253–58; K. M. Chin and M. S. Wolfe, "The Spread of *Erisiphe graminis* f. sp. *hordei* in Mixture of Barley Varieties," *Plant Pathology* 33 (1984): 89–100; M. S. Wolfe, "The Current Status and Prospects of Multiline Cultivars and Variety Mixtures for Resistance," *Annual Review of Phytopathology* 23 (1985): 251–73.

32. Jeremy J. Burdon and G. A. Chilvers, "The Effect of Barley Mildew on Barley and Wheat Competition in Mixtures," *Australian Journal of Botany* 25 (1977): 59–65; Christopher C. Mundt and J. A. Browning, "Development of Crown Rust Epidemics in Genetically Diverse Oat Populations: Effect of Genotype Unit Area," *Phytopathology* 75 (1985): 607–10; Christoper C. Mundt and K. J. Leonard, "Effect of Host Genotype Unit Area on Epidemic Development of Crown Rust Following Focal and General Inoculations of Mixtures of Immune and Susceptible Host Plants," *Phytopathology* 75 (1985): 1141–45; Chin and Wolfe, "The Spread of *Erisiphe graminis* f. sp. *hordei*."

33. The history of the search for a perennial agriculture at The Land Institute has been summarized by Wes Jackson in "Agriculture with Nature as Analogy," in Charles A. Francis, Cornelia Butler Flora, and Larry D. King, eds., *Sustainable Agriculture in Temperate Zones* (New York: John Wiley and Sons, 1990), 381–422.

34. Lynn Bates, Marty Bender, and Wes Jackson, "Eastern Gama Grass: Seed Structure and Protein Quality," *Cereal Chemistry* 58 (1981): 138–41; Tracy J. Bargman et al., "Compositional and Nutritional Evaluation of Eastern Gamagrass (*Tripsacum dactyloides* (L). L.), a Perennial Relative of Maize (*Zea mays* L.)," *Lebensmittel—Wissenschaft und Technologie* 22 (1989): 208–12.

35. Chet L. Dewald and R. S. Dayton, "A Prolific Sex Form Variant of Eastern Gamagrass," *Phytologia* 57 (1985): 156; Chet L. Dewald et al., "Morphology, Inheritance, and Evolutionary Significance of Sex Reversal in *Tripsacum dactyloides* (Poaceae)," *American Journal of Botany* 74 (1987): 1055–59.

36. Jon Piper et al., "Seed Yield and Quality Comparison of Herbaceous Perennials and Annual Crops," in Patricia Allen and Debra Van Dusen, eds., *Global Perspectives on Agroecology and Sustainable Agricultural Systems* (Santa Cruz: Agroecology Program, University of California, 1988), 715–19.

37. Melissa Luckow in personal communication with authors.

38. L. T. Leonard, "Lack of Nodule Formation in a Subfamily of the Leg-

uminosae," *Soil Science* 20 (1925): 165–67. Jon Piper has never seen nodulation of *C. marilandica* in field- or greenhouse-grown plants inoculated with *Rhizobium*.

39. T. Thompson, D. Zimmerman, and C. Rogers, "Wild *Helianthus* as a Genetic Resource," *Field Crops Research* 4 (1981): 333–43.

40. This study has been summarized in a submitted manuscript by Jon K. Piper et al., "Community Structure of Tallgrass Prairie Across a Productivity Gradient in Wet and Dry Years."

41. Jon K. Piper and Douglas Towne, "Multiple Year Patterns of Seed Yield in Five Herbaceous Perennials," *The Land Institute Research Report* 5 (1988): 14–18.

42. Brooks Anderson, "Progress Toward Breeding a Perennial Grain Sorghum," *The Land Institute Research Report* 6 (1989): 16–19.

43. Tamara Kraus, "Evaluation of Overwintering Capabilities in Sorghum Crosses: *Sorghum bicolor* × *Sorghum halepense*," *The Land Institute Research Report* 7 (1990): 31–33.

44. This experiment is detailed in Paul J. Muto, "Evaluation of Illinois Bundleflower and Eastern Gamagrass Germplasm in Monocultures and Bicultures," *The Land Institute Research Report* 7 (1990).

45. Details of this study are included in a submitted manuscript by Jon K. Piper, Elizabeth Gibans, and Bruce K. Kendall, "Growth Patterns and Seed Yields of Three Perennial Seed Crops in Polyculture."

46. Lois Braun, "Overyielding in a Perennial Legume Biculture," *The Land Institute Research Report Supplement* 2 (1985): 4–7.

47. Muto, "Evaluation of Illinois bundleflower and eastern gamagrass germplasm."

48. Peter A. Kulakow, Laura L. Benson, and Jake G. Vail, "Prospects for Domesticating Illinois Bundleflower," in J. Janick and J. E. Simon, eds., *Advances in New Crops* (Portland, Oreg.: Timber Press, 1990), 168–71.

49. S. M. Lofton, "Nitrogen Fixation in Selected Prairie Legumes as Related to Succession (Ph.D. diss., Oklahoma State University, 1976); W. C. Lindemann, G. W. Randall, and G. E. Ham, "Tillage Effects on Soybean Nodulation, $N_2(C_2H_4)$ Fixation, and Seed Yield," *Agronomy Journal* 74 (1982): 1067–70.

50. This study is summarized in a submitted manuscript by Jon K. Piper, "Patterns of Soil Moisture and Nutrient Change in Plots of Three Perennial Crops."

51. Piper, Gibans, and Kendall, "Growth Patterns and Seed Yield of Three Perennial Seed Crops."

52. Mary Handley of The Land Institute and Robert Bowden and Dallas Siefers of Kansas State University are presently working to identify the organisms that cause foliar diseases in eastern gamagrass.

53. Mary K. Handley et al., "Impact of Two Foliar Diseases on Growth and

Yield of Eastern Gamagrass," in *Eastern Gamagrass Conference Proceedings* (Poteau, Okla.: Kerr Center for Sustainable Agriculture, 1990), 31–39.

CHAPTER 6

1. The Asilomar Declaration for Sustainable Agriculture came out of the 1990 Ecological Farming Conference held at Asilomar, California. It was published in full in a number of sources, including *The Land Report* 37 (Spring 1990): 16–17.
2. Frances Moore Lappé and Joseph Collins, *World Hunger: Twelve Myths* (New York: Food First and Grove Press, 1986), 64.
3. David Barkin, Rosemary L. Batt, and Billie R. DeWalt, *Food Crops vs. Feed Crops: Global Substitution of Grains in Production* (Boulder: Lynne Rienner Publishers, 1990), 83.
4. Sylvan Wittwer, in "Research and Technology Needs for the Twenty-First Century," in M. S. Swaminathan and S. K. Sinha, eds., *Global Aspects of Food Production* (Oxford: Tycooly International), disagrees with this conclusion, stating that food acceptability is a major hurdle for crop improvement.
5. For descriptions of these innovators' systems, see Richard Thompson, Sharon Thompson, and Derrick Exner, "Case Study: A Resource-Efficient Farm with Livestock," pp. 263–80, and Rebecca W. Andrews, Steven E. Peters, Rhonda R. Janks, and Warren W. Sahs, "Converting to Sustainable Farming Systems," pp. 281–314, both in Charles A. Francis, Cornelia Butler Flora, and Larry D. King, eds., *Sustainable Agriculture in Temperate Zones* (New York: John Wiley and Sons, 1990).
6. National Research Council, *Alternative Agriculture* (Washington, D.C.: National Academy Press, 1989), 38.
7. Chuck Hassebrook, "Developing a Socially Sustainable Agriculture," *American Journal of Alternative Agriculture* 5 (1990): 96.
8. National Research Council, *Alternative Agriculture*, 78.
9. John Vandermeer, on pages 23 through 25 of *The Ecology of Intercropping* (Cambridge: Cambridge University Press, 1989), gives a thorough explanation of how to measure the value of intercrops based on various criteria, including monetary and dietary value of crops, and risk reduction.
10. For more concrete ideas, see the last chapter of *Family Farming* (Lincoln: University of Nebraska Press, 1988), where Marty Strange outlines new farm policies, tax laws, and loan programs aimed at encouraging entry into farming, protecting competition by limiting excessive competition, favoring the have-nots to redress inequities among farmers, and providing guidance for technology and land-use decisions that would protect the future common good.

11. Land Stewardship Project, "The 1990 Farm Bill: Environment Wins, Family Farmers Lose," *Land Stewardship Update* 1 (3: Winter 1990).

12. For accounts of these activities, see various issues of the Land Stewardship Project publication *The Land Stewardship Letter*, available from The Land Stewardship Project, 14758 Ostlund Trail North, Marine, MN 55047.

13. For more information on CSAs, see Greg Bowman, "Farms for Members Only," *New Farm* 13 (January 1991): 16–19.

14. See Chapter 2 for a discussion of publication pressures in research institutions.

15. Miguel Altieri, "Why Study Traditional Agriculture?" in C. Ronald Carroll, John H. Vandermeer, and Peter Rosset, eds., *Agroecology* (New York: McGraw-Hill, 1990), 560–61.

16. Richard Ness, "A Paradigm Shift for Extension: On-farm Research in Sustainable Farming Practices," *Land Stewardship Letter* 7 (4): 13.

17. William Lockeretz, "Establishing the Proper Role for On-farm Research," *American Journal of Alternative Agriculture* 2 (1987): 132–36.

18. Noted ecologist J. Stan Rowe, in a speech entitled "Land and People: An Ecological Perspective," suggests that we use the term *ecosphere* instead of the more widely used *biosphere* to get away from the human bias toward biological phenomena as the only things that matter. See *The Land Report No. 33* (Summer 1988): 10–15, for a full text of the speech.

19. Mark Leoschke of Iowa Natural Heritage Program in personal communication with Judith Soule.

20. John Buhnerkempe of Illinois Natural Heritage Database in personal communication with Judith Soule.

21. For example, see J. Emory, "Brazil's Pantanal, 'That Land of Wonderful Inundation,'" *Nature Conservancy* 40 (September–October 1990): 6–15.

22. Philip Fearnside, "Extractive Reserves in Brazilian Amazonia," *BioScience* 39 (1989): 387.

23. Executive Summary of The Nature Conservancy's Strategic Plan for the 1990s.

24. Doug Romig, "Sustainable Agriculture: Avoiding the Definition Trap," *The Land Report* 39 (1990): 15.

25. Henry Bauer, "Barriers Against Interdisciplinarity: Implications for Studies of Science, Technology, and Society (STS)," *Science, Technology, and Human Values* 15 (1990): 106–9.

26. Personal communication with Judith Soule.

27. Personal communication with Judith Soule.

28. Frederick Buttel, "Social Relations and the Growth of Modern Agriculture," in Carroll, Vandermeer, and Rosset, *Agroecology*, 138–39.

29. J. Mervis, "Panel Weighs Overhaul of NSF's Grant System," *The Scientist* 5 (1991): 1, 6–7, 12.

30. William Lacy et al., "Agricultural Biotechnology Research: Practices, Consequences, and Policy Recommendations," *Agriculture and Human Values* 5 (1988): 10.
31. David Pimentel et al., "Benefits and Risks of Genetic Engineering in Agriculture," *BioScience* 39 (1989): 606.
32. Lacy et al., "Agricultural Biotechnology Research," 8.
33. Lacy et al., "Agricultural Biotechnology Research," 7.
34. J. Sousa Silva, "The Contradictions of the Biorevolution for the Development of Agriculture in the Third World: Biotechnology and Capitalist Interest," *Agriculture and Human Values* 5 (1988): 64.
35. Lacy, et al., "Agricultural Biotechnology Research," 6, 10.
36. For example, see the chapter entitled "Beyond the Mongongo Tree: Good News about Conservation Tillage and the Environmental Trade-off" in Charles E. Little's book *Green Fields Forever* (Washington, D.C.: Island Press, 1987).
37. For example, see Pimentel et al., "Benefits and Risks of Genetic Engineering."
38. For examples of this problem, see Norman Ellstrand and Carol Hoffman, "Hybridization as an Avenue of Escape for Engineered Genes," *BioScience* 40(6): 438; see also Carol Hoffman, "Ecological Risks of Genetic Engineering of Crop Plants," in the same issue of *BioScience*, 434.
39. Mark Sagoff, "Biotechnology and the Environment: What Is at Risk? *Agriculture and Human Values* 5 (1988): 27.
40. See Lacy et al., "Agricultural Biotechnology Research," 10.
41. J. Sousa Silva, "Contradictions of the Biorevolution," 66.
42. Elisabet Sahtouris, *Gaia: The Human Journey from Chaos to Cosmos* (New York: Pocket Books, 1989), 9.
43. Rowe, "Land and People," 10.
44. This proposal was put forth by Wittwer, "Research and Technology," 96.
45. William Rees, in "The Ecology of Sustainable Development," *The Ecologist* 20 (1990): 18–23, gives some concrete suggestions on how to account for nonmonetary factors in economic terms.
46. Rowe, "Land and People," 11.
47. Wendell Berry, *Home Economics* (San Francisco: North Point Press, 1987), 60.
48. Wes Jackson, *Altars of Unhewn Stone* (San Francisco: North Point Press, 1987), 75.
49. David Pimentel et al., "A Cost-benefit Analysis of Pesticide Use in U.S. Food Production," *Pesticides* (Clifton, N.J.: Humana Press, 1979).
50. From the Asilomar Declaration for Sustainable Agriculture.
51. Neil Hamilton, "Don't Get Sued When You Spray," *New Farm* 12 (February 1990): 7.
52. Altieri, "Why Study Traditional Agriculture?" 562.

Bibliography

Adams, M. W., A. H. Ellingboe, and E. C. Rossineau. "Biological Uniformity and Disease Epidemics." *BioScience* 21 (1971): 1067–70.

Addiscott, Tom. "Farmers, Fertilisers and the Nitrate Flood." *New Scientist* 120 (1988): 50–54.

Ahring, Robert M. *The Management of Buffalograss Buchloe dactyloides (Nutt.) Engelm. for Seed Production in Oklahoma.* Oklahoma Agricultural Experiment Station Technical Bulletin T-109. Stillwater, Okla.: Oklahoma State University, 1964.

Alabama Agricultural Experiment Station. *Fifty-third Annual Report of the Alabama Agricultural Experiment Station of the Alabama Polytechnic Institute.* Auburn, Ala.: Alabama Polytechnic Institute, 1942.

Alderson, Lawrence. *The Chance to Survive: Rare Breeds in a Changing World.* London: Cameron and Taylor, in association with David and Charles, 1978.

Alexander, Helen M. "Spatial Patterns of Disease Introduced by *Fusarium moniliforme* var *subglutinans* in a Population of *Plantago lanceolata*." *Oecologia* 62 (1984): 141–43.

Alford, Harold G., and Mary P. Ferguson, eds. *Pesticides in Soil and Groundwater.* Berkeley: Division of Agricultural Sciences, University of California, 1983.

Alt, Klaus, C. Tim Osborn, and Don Colacicco. *Soil Erosion: What Effect on Agricultural Productivity?* Agriculture Information Bulletin No. 556. Washington, D.C.: U.S. Department of Agriculture, Economic Research Service, January 1989.

Altieri, Miguel A. "Agroecology: A New Research and Development Paradigm for World Agriculture." *Agriculture, Ecosystems and Environment* 27 (1989): 37–46.

———. "Why Study Traditional Agriculture?" In *Agroecology,* edited by C. Ronald Carroll, John H. Vandermeer, and Peter M. Rosset, 551–64. New York: McGraw-Hill, 1990.

Altieri, Miguel A., and Matthew Liebman. "Insect, Weed, and Plant Disease Management in Multiple Cropping Systems." In *Multiple Cropping Systems,* edited by Charles A. Francis, 183–218. New York: Macmillan, 1986.

Anderson, Brooks. "Progress Toward Breeding a Perennial Grain Sorghum." *The Land Institute Research Report* 6 (1989): 16–19.

Anderson, D. W., and David C. Coleman. "The Dynamics of Organic Matter in Grassland Soils." *Journal of Soil and Water Conservation* 40 (1985): 211–16.

Andow, David A. "Plant Diversity and Insect Populations in Experimental Agroecosystems: Interactions Among Beans, Insects and Weeds." Ph.D. diss., Cornell University, 1983.

Andow, David A., and Peter M. Rosset. "Integrated Pest Management." In *Agroecology,* edited by C. Ronald Carroll, John H. Vandermeer, and Peter M. Rosset, , 413–39. New York: McGraw-Hill, 1990.

Ashworth, William. *Nor Any Drop to Drink.* New York: Summit Books, 1982.

Augspurger, Carol K. "Seed Dispersal of the Tropical Tree, *Platypodium elegans,* and the Escape of its Seedlings from Fungal Pathogens." *Journal of Ecology* 71 (1983): 759–71.

———. "Seedling Survival of Tropical Tree Seedlings: Interactions of Dispersal Distance, Light-gaps and Pathogens. *Ecology* 65 (1984): 1705–12.

Badger, Curtis J., "Eastern Shore Gold." *Nature Conservancy* 40 (July-August 1990): 6–15.

Baker, J. L., and J. M. Laflen. "Water Quality Consequences of Conservation Tillage." *Journal of Soil and Water Conservation* 38 (1983): 186–93.

Baker, J. L., R. S. Kanwar, and T. A. Austin. "Impact of Agricultural Drainage Wells on Groundwater Quality." *Journal of Soil and Water Conservation* 40 (1985): 516–20.

Banker, Dave. "Agriculture's Financial Profile: A Look Behind the Aggregates." *Farmline* (December-January 1987): 6–7.

Banker, Ellen. "How Healthy are Rural Banks?" *Farmline* (August 1986): 10–11.

Bargman, Tracy J., Grace D. Hanners, Robert Becker, Robin M. Saunders, and John H. Rupnow. "Compositional and Nutritional Evaluation of Eastern Gamagrass (*Tripsacum dactyloides* (L). L.), a Perennial Relative of Maize (*Zea mays* L.)." *Lebensmittel—Wissenschaft und Technologie* 22 (1989): 208–12.

Barkin, David, Rosemary L. Batt, and Billie R. DeWalt. *Food Crops vs. Feed Crops: Global Substitution of Grains in Production.* Boulder: Lynne Rienner Publishers, 1990.

Barnett, Frank L., and Gary L. Posler. "Performance of Cool-season Perennial Grasses in Pure Stand and in Mixtures with Legumes." *Agronomy Journal* 75 (1983): 582–86.

Barrett, J. A. "The Evolutionary Consequences of Monoculture." In *Genetic Consequences of Man-made Change*, edited by J. A. Bishop and L. M. Cook, 209–48. London: Academic Press, 1981.

Bartholomew, W. V., W. D. Shrader, and A. J. Englehorn. "Nitrogen Changes Attending Various Crop Rotations on Clarion-Webster Soils in Iowa." *Agronomy Journal* 49 (1957): 415–18.

Bates, Lynn, Marty Bender, and Wes Jackson. "Eastern Gama Grass: Seed Structure and Protein Quality." *Cereal Chemistry* 58 (1981): 138–41.

Batie, Sandra S. *Soil Erosion: Crisis in America's Croplands?* Washington, D.C.: The Conservation Foundation, 1983.

Bauer, Henry H. "Barriers Against Interdisciplinarity: Implications for Studies of Science, Technology, and Society (STS)." *Science, Technology, and Human Values* 15 (1990): 105–19.

Becker, Robert, Peggy Wagoner, Grace D. Hanners, and Robin M. Saunders. "Compositional, Nutritional, and Functional Evaluation of Intermediate Wheatgrass (*Thinopyrum intermedium*)." *Journal of Food Processing and Preservation* 15 (1991): 63–77.

Beer, John. "Litter Production and Nutrient Cycling in Coffee (*Coffea arabica*) or Cacao (*Theobroma cacao*) Plantations with Shade Trees." *Agroforestry Systems* 7 (1988): 103–14.

Bender, Marty. "Industrial Versus Biological Traction on the Farm." In *Meeting the Expectations of the Land*, edited by Wes Jackson, Wendell Berry, and Bruce Colman, 87–105. San Francisco: North Point Press, 1984.

Benzioni, A., and R. L. Dunstone. "Jojoba: Adaptation to Environmental Stress and Implications for Domestication." *Quarterly Review of Biology* 61 (1986): 177–99.

Berish, Cory W., and John J. Ewel. "Root Development in Simple and Complex Tropical Successional Ecosystems." *Plant and Soil* 106 (1988): 73–84.

Berry, Wendell. *The Unsettling of America: Culture and Agriculture.* San Francisco: Sierra Club Books, 1977.

———. *The Gift of Good Land.* San Francisco: North Point Press, 1981.

———. *Home Economics.* San Francisco: North Point Press, 1987.

Bertrand, A. "Biological Diversity: Basic for Agricultural Success." In *Proceedings of the U.S. Strategy Conference on Biological Diversity*, November 16–18, 1981. Publication 9262, 28–33. Washington, D.C.: Department of State, 1982.

Blair, Aaron, and Terry L. Thomas. "Leukemia Among Nebraska Farmers: A Death Certificate Study." *American Journal of Epidemiology* 110 (1979): 264–73.

Blank, R. R., and M. A. Fosberg. "Cultivated and Adjacent Virgin Soils in Northcentral South Dakota: I. Chemical and Physical Comparisons." *Soil Science Society of America Journal* 53 (1989): 1484–90.

Blumberg, A. Y., and D. A. Crossley, Jr. "Comparison of Soil Surface Arthropod Populations in Conservation Tillage, No-tillage, and Old-field Systems." *Agro-Ecosystems* 8 (1983): 247–53.

Bodrov, M. S. "Hybridization between Wheat and *Elymus*." In *Wide Hybridization in Plants*, edited by N. V. Tsitsin, 238–41. Moscow: Academy of Sciences of the U.S.S.R. All-Union Academy of Agricultural Sciences, 1960. (Published for the National Science Foundation, Washington, D.C., by the Israeli Program for Scientific Translations, Jerusalem, 1962.)

Bohrer, V. L. "The Prehistoric and Historic Role of the Cool-season Grasses in the Southwest." *Economic Botany* 29 (1975): 199–207.

Bormann, F. Herbert. "Air Pollution and Forests: An Ecosystem Perspective." *BioScience* 35 (1985): 434–41.

Bormann, F. Herbert, and Gene E. Likens. *Pattern and Process in a Forested Ecosystem*. New York: Springer-Verlag, 1979.

Botkin, Daniel B. *Discordant Harmonies*. New York: Oxford University Press, 1990.

Bowman, Greg. "Farms for Members Only." *New Farm* 13 (January 1991): 16–19.

Brady, N. C. "Welcome and Introduction." In *Proceedings of the U.S. Strategy Conference on Biological Diversity*, November 16–18, 1981. Publication 9262, 11–13. Washington, D.C.: Department of State, 1982.

Braun, Lois. "Overyielding in a Perennial Legume Biculture." *The Land Institute Research Report Supplement* 2 (1985): 4–7.

Briggle, L. W. "Introduction to Energy Use in Wheat Production." In *Handbook of Energy Utilization in Agriculture*, edited by David Pimentel, 109–16. Boca Raton, Fla.: CRC Press, 1980.

Brohan, Mark. "Tight Times for Rural Governments." *Farmline* (November 1986): 8–9.

Brophy, L. S., G. H. Heichel, and M. P. Russelle. "Nitrogen Transfer from Forage Legumes to Grass in a Systematic Planting Design." *Crop Science* 27 (1987): 753–58.

Brown, Anthony W. A. *Ecology of Pesticides*. New York: John Wiley and Sons, 1978.

Brown, H. R. "Growth and Seed Yields of Native Prairie Plants in Various Habitats of the Mixed Prairie." *Transactions of the Kansas Academy of Science* 46 (1943): 87–99.

Brown, Lester. "Conserving Soils." In *State of the World 1984*, 53–73. Washington, D.C.: Worldwatch Institute, 1984.

———. "The Vulnerability of Oil-based Farming." *World Watch* 1(2) (1988): 24–29.

Brown, N. J. "Biological Diversity: The Global Challenge." In *Proceedings of the U.S. Strategy Conference on Biological Diversity,* November 16–18, 1981. Publication 9262, 22–26. Washington, D.C.: Department of State, 1982.

Brown, R. H., and G. T. Byrd. "Yield and Botanical Composition of Alfalfa-Bermudagrass Mixtures." *Agronomy Journal* 82 (1990): 1074–79.

Brust, Gerald E., Benjamin R. Stinner, and David A. McCartney. "Predator Activity and Predation in Corn Agroecosystems." *Environmental Entomology* 15 (1986): 1017–21.

Bull, David A. *A Growing Problem: Pesticides and the Third World Poor.* Oxford: Oxfam, 1982.

Buranday, R. P., and R. S. Raros. "Effects of Cabbage-Tomato Intercropping on the Incidence and Oviposition of the Diamond-back Moth, *Plutella xylostella* (L.)." *Philippines Entomology* 2 (1975): 369–74.

Burdon, Jeremy J. "Mechanisms of Effective Disease Control in Heterogeneous Plant Populations—An Ecologist's View." In *Plant Disease Epidemiology,* edited by P. R. Scott and A. Bainbridge, 193–200. New York: Blackwell Scientific Publishers, 1978.

Burdon, Jeremy J. *Diseases and Plant Population Biology.* Cambridge: Cambridge University Press, 1987.

Burdon, Jeremy J., and G. A. Chilvers. "The Effect of Barley Mildew on Barley and Wheat Competition in Mixtures." *Australian Journal of Botany* 25 (1977): 59–65.

———. "Host Density as a Factor in Plant Disease Ecology." *Annual Review of Phytopathology* 20 (1982): 143–66.

Burdon, Jeremy J., and D. R. Marshall. "Inter- and Intra-specific Diversity in the Disease-response of *Glycine* Species to the Leaf Rust Fungus *Phakopsora pachyrhizi.*" *Journal of Ecology* 69 (1981): 381–90.

Burdon, Jeremy J., J. D. Oates, and D. R. Marshall. "Interactions between *Avena* and *Puccinia* Species. I. The Wild Hosts: *Avena barbata* Pott ex Link, *A. fatus* L., *A. ludoviciana* Durieu." *Journal of Applied Ecology* 20 (1983): 571–84.

Burdon, Jeremy J., and R. Whitbread. "Rates of Increase of Barley Mildew in Mixed Stands of Wheat and Barley." *Journal of Applied Ecology* 16 (1979): 253–58.

Burkhardt, Jeffrey. "Biotechnology, Ethics, and the Structure of Agriculture." *Agriculture and Human Values* 5 (1988): 53–60.

Burmeister, Leon F., George D. Everett, Stephanie F. Van Lier, and Peter Isacson. "Selected Cancer Mortality and Farm Practices In Iowa." *American Journal of Epidemiology* 118 (1983): 72–77.

Burritt, Brad. "Leymus: A Plant with a History of Human Use." *The Land Report* 28 (1986): 10–12.

Busch, Lawrence, and William B. Lacy. *Science, Agriculture, and the Politics of Research.* Boulder: Westview Press, 1983.

————. "Agricultural Research in Developed Nations." In *Agroecology*, edited by C. Ronald Carroll, John H. Vandermeer, and Peter M. Rosset, 565–81. New York: McGraw-Hill, 1990.

Buttel, Frederick H. "Social Relations and the Growth of Modern Agriculture." In *Agroecology*, edited by C. Ronald Carroll, John H. Vandermeer, and Peter M. Rosset, 113–46. New York: McGraw-Hill, 1990.

Buttel, Frederick H., and I. Garth Youngberg. "Implications of Biotechnology for the Development of Sustainable Agricultural Systems." In *Environmentally Sound Agriculture*, edited by W. Lockeretz, 377–400. New York: Praeger, 1983.

Cantor, Kenneth P. "Farming and Mortality from Non-Hodgkin's Lymphoma: A Case-control Study." *International Journal of Cancer* 29 (1982): 239–47.

Cantor, Kenneth P., and Aaron Blair. "Farming and Mortality from Multiple Myeloma: A Case-control Study with the Use of Death Certificates." *Journal of the National Cancer Institute* 72 (1984): 251–55.

Carroll, C. Ronald. "The Interface Between Natural Areas and Agroecosystems." In *Agroecology*, edited by C. Ronald Carroll, John H. Vandermeer, and Peter M. Rosset, 365–84. New York: McGraw-Hill, 1990.

Carson, Rachel. *Silent Spring*. Boston: Houghton Mifflin, 1962.

Chaboussou, Francis. "How Pesticides Increase Pests." *The Ecologist* 16(1) (1986): 29–35.

Chen, Hsiu-Hsiung, and A. Douglas Druliner. "Agricultural Chemical Contamination of Ground Water in Six Areas of the High Plains Aquifer, Nebraska." In *National Water Summary 1986—Hydrologic Events and Groundwater Quality*. U. S. Geological Survey Water-Supply Paper 2325, 103–108. Reston, Va.: U.S. Department of Interior, Geological Survey, for sale by Superintendent of Documents, U.S. Government Printing Office.

Chiarello, Nona, James C. Hickman, and Harold A. Mooney. "Endomycorrhizal Role for Interspecific Transfer of Phosphorus in a Community of Annual Plants." *Science* 217 (1982) 941–43.

Chin, K. M., and M. S. Wolfe. "The Spread of *Erysiphe graminis* f. sp. *hordei* in Mixture of Barley Varieties." *Plant Pathology* 33 (1984): 89–100.

Christie, P., E. I. Newman, and R. Campbell. "Grassland Species Can Influence the Abundance of Microbes on Each Other's Roots." *Nature* 250 (1974): 570–71.

Clark, Edwin H., II. "The Off-site Costs of Soil Erosion." *Journal of Soil and Water Conservation* 40 (1985): 19–26.

Clements, Frederick C. "Nature and Structure of the Climax." *Journal of Ecology* 24 (1936): 252–84.

Coaker, T. H. "Crop Pest Problems Resulting from Chemical Control." In *Origins of Pest, Parasite, Disease and Weed Problems*, edited by J. M. Cherrett and G. R. Sagar, 313–28. The Eighteenth Symposium of the British Ecological Society. Oxford: Blackwell Scientific Publications, 1977.

Cobb, F. W., Jr., G. W. Slaughter, D. L. Rowny, and C. J. DeMars. "Rate of Spread of *Ceratocystis wagneri* in Ponderosa Pine Stands in the Central Sierra Nevada." *Phytopathology* 72 (1982): 1359–62.

Cochrane, Willard W. *Farm Prices: Myth and Reality.* Minneapolis: University of Minnesota Press, 1958.

Coleman, David C., C. Vern Cole, and Edward T. Elliot. "Decomposition, Organic Matter Turnover, and Nutrient Dynamics in Agroecosystems." In *Agricultural Ecosystems,* edited by Richard Lowrance, Benjamin R. Stinner, and Garfield J. House, 83–104. New York: John Wiley and Sons, 1984.

Collins, Joseph. *Nicaragua: What Difference Could a Revolution Make?* New York: Food First and Grove Press, 1986.

Collins, Scott L. "Interaction of Disturbances in Tallgrass Prairie: A Field Experiment." *Ecology* 68 (1987): 1243–50.

Committee on Germplasm Resources. *Conservation of Germplasm Resources: An Imperative.* Washington, D.C.: National Academy of Sciences, 1978.

Connell, Joseph H. "Diversity in Tropical Rain Forests and Coral Reefs." *Science* 199 (1978): 1302–10.

———. "Diversity and the Coevolution of Competitors, or the Ghost of Competition Past." *Oikos* 35 (1980): 131–38.

Connell, Joseph H., and Ralph O. Slatyer. "Mechanisms of Succession in Natural Communities and Their Role in Community Stability and Organization." *American Naturalist* 111 (1977): 1119–44.

Conway, Gordon R. "The Future." In *Pesticide Resistance and World Food Production,* edited by Gordon R. Conway, 77–90. London: Imperial College Centre for Environmental Technology, 1982.

Cooper, H. W. *Producing and Harvesting Grass Seed in the Great Plains.* USDA Farmer's Bulletin No. 2112. 1957.

Council of Environmental Quality. *The Eleventh Annual Report of the Council of Environmental Quality.* Washington, D.C.: U.S. Government Printing Office, 1980.

Coye, Molly Joel. "The Health Effects of Agricultural Production: I. The Health of Agricultural Workers." *Journal of Public Health Policy* 6 (1985): 349–70.

———. "The Health Effects of Agricultural Production: II. The Health of the Community." *Journal of Public Health Policy* 7 (1986): 340–54.

Cramer, Craig. "The World Discovers Nature's Ag School." *New Farm* 11 (1989): 10–17.

———. "A 'Sustainable' Farm Bill." *New Farm* 13 (January 1991): 28–31.

Cronon, William. *Changes in the Land.* New York: Hill & Wang, 1983.

Crosby, E. A. "Private Sector Agricultural Research in the United States." In *Policy for Agricultural Research,* edited by V. W. Ruttan and C. E. Ray, 395–409. Boulder: Westview, 1987.

Dann, Paul. "Toward Self-sufficient Food Production." *Agricultural Science* (June-July 1988): 16–19.

Davidson, Osha. "The Rise of the Rural Ghetto." *The Nation* 14 June 1986, 820, 822.

Davis, Jeremy H. C., Jonathan N. Woolley, and Raul A. Moreno. "Multiple Cropping with Legumes and Starchy Roots." In *Multiple Cropping Systems,* edited by Charles A. Francis, 133–60. New York: Macmillan, 1986.

Dawkins, R., and J. Krebs. "Arms Races Between and Within Species." *Proceedings of the Royal Society of London Series B.* 205 (1979): 489–511.

Day, P. R. "The Genetic Base of Epidemics." In *Plant Disease: An Advanced Treatise. Vol. 2. How Disease Develops in Populations,* edited by James G. Horsfall and Ellis B. Cowling, 263–85. New York: Academic Press, 1978.

Debach, Paul. *Biological Control by Natural Enemies.* London: Cambridge University Press, 1974.

Derzhavin, A. I. "The Theory and Practice of Producing Perennial Rye Varieties." In *Wide Hybridization in Plants,* edited by N. V. Tsitsin, 143–52. Moscow: Academy of Sciences of the U.S.S.R. All-Union Academy of Agricultural Sciences, 1960. (Published for the National Science Foundation, Washington, D.C., by the Israeli Program for Scientific Translations, Jerusalem, 1962.)

Dewald, C. L., B. L. Burson, J. M. J. de Wet, and J. R. Harlan. "Morphology, Inheritance, and Evolutionary Significance of Sex Reversal in *Tripsacum dactyloides* (Poaceae)." *American Journal of Botany* 74 (1987): 1055–59.

Dewald, C. L., and R. S. Dayton. "A Prolific Sex Form Variant of Eastern Gamagrass." *Phytologia* 57 (1985): 156.

Dickason, Clifford. "Improved Estimates of Groundwater-mining Acreage." *Journal of Soil and Water Conservation* 43: (1988): 239–56.

Dinoor, A., and N. Eshed. "The Role and Importance of Pathogens in Natural Plant Communities." *Annual Review of Phytopathology* 22 (1984): 443–66.

Doebley, John F., "'Seeds' of Wild Grasses: A Major Food of Southwestern Indians." *Economic Botany* 38 (1984): 52–64.

Door, Robert. "Insurers Own More Land, Group Says." *Sunday World-Herald,* 26 July 1987.

Drolsom, P. N., and D. Smith. "Adapting Species for Forage Mixtures." In *Multiple Cropping,* edited by R. I. Papendick, P. A. Sanchez, and G. B. Triplett, 223–34. ASA Special Publication No. 27. Madison, Wisc.: American Society of Agronomy, Crop Science Society of America, and Soil Science Society of America, 1976.

Dundon, Stanislaus J. "Hidden Obstacles to Creativity in Agricultural Science." *Twentieth Century Agricultural Science* n.s., vol. 8 (1983): 105–26.

Ecological Farming Conference. The Asilomar Declaration for Sustainable Agriculture. *The Land Report* 37 (Spring 1990): 16–17.

"EDB's Long-lasting Legacy." *Science News* 127 (1985): 313.

Edwards, C. A., and J. F. Lofty. "The Influence of Cultivation on Animal Populations." In *Progress in Soil Biology,* edited by J. Vanek, 399–408. Prague: Academia Publications, 1975.

Eisner, Thomas, and Jacob Gould Schurman. "Chemicals, Genes, and the Loss of Species." *Nature Conservancy News* 33(6) (1983): 23–24.

Eissenstat, D. M. "A Comparison of Phosphorus and Nitrogen Transfer Between Plants of Different Phosphorus Status." *Oecologia* 82 (1990): 342–47.

El-Ashry, Mohamed T., Jan van Schilfgaarde, and Susan Schiffman. "Salinity Pollution from Irrigated Agriculture." *Journal of Soil and Water Conservation* 40 (1985): 48–52.

Ellstrand, Norman C., and Carol A. Hoffman. "Hybridization as an Avenue of Escape for Engineered Genes." *BioScience* 40 (1990): 438–42.

El-Sebae, A. H. "Global Impact of Pesticides." *Environment Abstracts Annual* 18 (1988): A47–A50.

Emory, Jerry. "Brazil's Pantanal, 'That Land of Wonderful Inundation.'" *Nature Conservancy* 40 (September–October 1990): 6–15.

"E.P.A. Finds Pesticides in Water of Thirty-eight States." *New York Times,* 14 December 1988.

Ewel, John, Cory Berish, Becky Brown, Norman Price, and James Raich. "Slash and Burn Impacts on a Costa Rican Wet Forest Site." *Ecology* 62 (1981): 816–29.

Experiment Station Committee on Organization and Policy (ESCOP). "Research Initiatives: A Midterm Update of the Research Agenda for the State Agricultural Experiment Stations." College Station, Tex.: Texas Agricultural Experiment Station and Texas A & M University, 1988.

Fairchild, Deborah M. "A National Assessment of Ground Water Contamination from Pesticides and Fertilizers." In *Ground Water Quality and Agricultural Practices,* edited by D. M. Fairchild, 273–94. Chelsea, Mich.: Lewis Publishers, 1987.

Farley, Dixie. "Setting Safe Limits on Pesticide Residues." *FDA Consumer* 22(8) (1988): 8–11.

Fearnside, Philip M. "Extractive Reserves in Brazilian Amazonia." *Bioscience* 39 (1989) 387–93.

Feeny, Paul. "Plant Apparency and Chemical Defense." *Recent Advances in Phytochemistry* 10 (1976): 1–39.

Fenner, Louise. "A Hard Look at What We're Eating." *FDA Consumer* 18(3) (1984): 8–13.

Ferguson, Bruce K. "Wither Water? The Fragile Future of the World's Most Important Resource." *The Futurist* 17 (April 1983): 29–47.

Flannery, Kent V. "The Origins of Agriculture." *Annual Review of Anthropology* 2 (1973): 271–310.

Fleming, Malcolm H. "Agricultural Chemicals in Ground Water: Preventing Contamination by Removing Barriers against Low-input Farm Management." *American Journal of Alternative Agriculture* 2 (1987): 124–30.

Follett, R. F., and G. A. Peterson. "Surface Soil Nutrient Distribution as Affected by Wheat-fallow Tillage Systems." *Soil Science Society of America Journal* 52 (1988): 141–47.

Food and Agriculture Organization. *Report of the Second Session of the FAO Panel of Experts on Pest Resistance to Pesticides and Crop Loss Assessment.* Rome: Food and Agriculture Organization of the United Nations, 1979.

Forsyth, C., and N. A. C. Brown. "Germination of the Dimorphic Fruits of *Bidens pilosa* L." *New Phytologist* 90 (1982): 151–64.

Foster, David R. "Distrubance History, Community Organization and Vegetation Dynamics of the Old-growth Pisgah Forest, Southwestern New Hampshire, U.S.A." *Journal of Ecology* 76 (1988): 105–34.

Foulke, J. "Off-farm Income Vital for Many Farmers." *Farmline* (June 1988): 8–9.

Francis, Charles A., ed. *Multiple Cropping Systems.* New York: Macmillan, 1986.

Francis, Charles A. "Sustaining Agriculture and Development: Challenges for the Future." *American Journal of Alternative Agriculture* 4 (1989): 98–99.

Francis, Charles A., Cornelia Butler Flora, and Larry D. King, eds. *Sustainable Agriculture in Temperate Zones.* New York: John Wiley and Sons, 1990.

Francis, Charles A., C. A. Flor, and S. R. Temple. "Adapting Varieties for Intercropping Systems in the Tropics." In *Multiple Cropping,* edited by R. I. Papendick, P. A. Sanchez, and G. B. Triplett, 235–53. ASA Special Publication No. 27. Madison, Wisc.: American Society of Agronomy, Crop Science Society of America, and Soil Science Society of America, 1976.

Fredrick, K. D., and J. Hanson. *Water for Western Agriculture.* Washington, D.C.: Resources for the Future and Johns Hopkins University Press, 1982.

Freedman, B., and T. C. Hutchinson. "Long Term Effects of Smelter Pollution at Sudbury, Ontario, on Surrounding Forest Communities." *Canadian Journal of Botany* 58 (1980): 2123–40.

Fruhling, Larry. "Please Don't Drink the Water." *The Progressive* 50 (1986): 31–33.

General Accounting Office. *Pesticides: Better Sampling and Enforcement Need on Imported Food.* GAO Report RCED-86–219, September 26, 1986.

————. *Pesticides: Need to Enhance FDA's Ability to Protect the Public from Illegal Residues.* GAO Report RCED-87–7, October 27, 1986.

Georghiou, George P. "The Magnitude of the Resistance Problem." In *Pesticide Resistance: Strategies and Tactics for Management,* 14–43. National Resource Council Committee on Strategies for the Management of Pesticide Resistant Pest Populations. Washington, D.C.: National Academy Press, 1986.

Gernes, Mark. "Effect of *Helianthus maximilianii* Density on Biomass of Naturally Occurring Weeds." *The Land Institute Research Report Supplement* 3 (1986): 39–42.

Gilliom, Robert J., Richard B. Alexander, and Richard A. Smith. *Pesticides in the Nation's Rivers, 1975–1980, and Implications for Future Monitoring.* U.S. Geological Survey Water-Supply Paper 2271. Reston, Va.: Department of Interior, U.S. Geological Survey, 1987.

Gliessman, Stephen R. "An Agroecological Approach to Sustainable Agriculture." In *Meeting the Expectations of the Land*, edited by Wes Jackson, Wendell Berry, and Bruce Colman, 160–71. San Francisco: North Point Press, 1984.

Goodman, P. J., and M. Collison. "Varietal Differences in Uptake of ^{32}P Labelled Phosphate in Clover Plus Ryegrass Swards and Monocultures." *Annals of Applied Biology* 100 (1982): 559–65.

Gore Peter H., Paul Davis, and Thomas Sleight. "Pesticides and Peasants: They Don't Mix." *The Rural Sociologist* 3 (1983): 98–101.

Great Plains Flora Association. *Flora of the Great Plains*. Lawrence: University Press of Kansas, 1986.

Grieshop, James I., with David Winter. "Communication for Safety's Sake: Visual Communication Materials for Pesticide Users in Latin America." *Tropical Pest Management* 34 (1988): 249–62.

Groffman, Peter M., Paul F. Hendrix, and D. A. Crossley, Jr. "Nitrogen Dynamics in Conventional and No-tillage Agroecosystems with Fertilizer or Legume Nitrogen Inputs." *Plant and Soil* 97 (1987): 315–22.

Gupta, Y. P. "Pesticide Misuse in India." *The Ecologist* 16(1) (1986): 36–39.

Haas, H. J., C. E. Evans, and E. F. Miles. *Nitrogen and Carbon Changes in Great Plains Soils as Influenced by Cropping and Soil Treatments*. USDA Technical Bulletin 1164. Washington, D.C.: U.S. Government Printing Office, 1957.

Hadwiger, Don F. *The Politics of Agricultural Research*. Lincoln: University of Nebraska Press, 1982.

Hallberg, George R. "From Hoes to Herbicides: Agriculture and Groundwater Quality." *Journal of Soil and Water Conservation* 41 (1986): 357–64.

Hamilton, Neil D. "Don't Get Sued When You Spray." *New Farm* 12 (February 1990): 7–8.

Handley, Mary K., Peter A. Kulakow, James Henson, and Chet L. Dewald. "Impact of Two Foliar Diseases on Growth and Yield of Eastern Gamagrass." In *Eastern Gamagrass Conference Proceedings*, 31–39. Poteau, Okla.: Kerr Center for Sustainable Agriculture, 1990.

Harl, Neil E. "The People and the Institutions: An Economic Assessment." *Farm Foundation: Increasing Understanding of Public Problems and Policies* (1986): 71–89.

Harlan, Jack R. "Diseases as a Factor in Plant Evolution." *Anual Review of Phytopathology* 14 (1976): 31–51.

———. "Wild Grass Seeds as Food Sources in the Sahara and Sub-Sahara." *Sahara* (2/1989): 69–74.

Harlan, Jack R., and Robert M. Ahring. *Caddo Switchgrass*. Oklahoma Agricultural Experiment Station Bulletin B-516. Stillwater, Okla.: Oklahoma State University, 1958.

Harlan, Jack R., and Daniel Zohary. "Distribution of Wild Wheats and Barley." *Science* 153 (1966): 1074–80.

Harper, John L. *Population Biology of Plants*. London: Academic Press, 1977.

Harrington, Dave. "How the Crisis Will—and Won't—Change U.S. Agriculture." *Farmline* (April 1987): 10–11.

Harris, W. F., P. Sollins, N. T. Edwards, B. E. Dinger, and H. H. Shugart. "Analysis of Carbon Flow and Productivity in a Temperate Deciduous Forest Ecosystem." In *Productivity of World Ecosystems*, 116–22. Washington, D.C.: National Academy of Sciences, 1975.

Harrison, Jack. "U. S. Farm Sector Weathers Storm, Builds Strength." *Farmline* (December 1989–January 1990): 4–8.

Hart, Robert D. "A Natural Ecosystem Analog Approach to the Design of a Successional Crop System for Tropical Forest Environments." *Tropical Succession (Biotropica* Supplement) 12 (1980): 73–82.

Harwood, Richard R., and E. C. Price. "Multiple Cropping in Tropical Asia." In *Multiple Cropping*, edited by R. I. Papendick, P. A. Sanchez, and G. B. Triplett, 11–40. ASA Special Publication No. 27. Madison, Wisc.: ASA-CSSA-SSSA, 1976.

Hassebrook, Chuck. "Developing a Socially Sustainable Agriculture." *American Journal of Alternative Agriculture* 5 (1990): 50, 96.

Heffernan, William D., and Judith Bortner Heffernan. "Impact of the Farm Crisis on Rural Families and Communities." *Rural Sociologist* 6 (1986): 160–69.

Hendrix, Paul F., Robert W. Parmelee, D. A. Crossley, Jr., David C. Coleman, Eugene P. Odum, and P. M. Groffman. "Detritus Food Webs in Conventional and No-tillage Agroecosystems." *BioScience* 36 (1986): 374–80.

Hetrick, Barbara A. Daniels, D. Gerschschefske Kitt, and D. Thompson Wilson. "Mycorrhizal Dependence and Growth Habit of Warm-season and Cool-season Tallgrass Prairie Plants." *Canadian Journal of Botany* 66 (1988): 1376–80.

Hetrick, Barbara A. Daniels, G. T. Wilson, and David C. Hartnett. "Relationship between Mycorrhizal Dependence and Competitive Ability of Two Tallgrass Prairie Grasses." *Canadian Journal of Botany* 67 (1989): 2608–15.

Hightower, Jim. "Hard Tomatoes, Hard Times: The Failure of the Land Grant College Complex." In *Radical Agriculture*, edited by R. Merrill, 87–110. New York: Harper and Row, 1976.

Hill, A. R. "Nitrate Distribution in the Ground Water of the Alliston Region of Ontario, Canada." *Ground Water* 20 (1982): 696–702.

Hinkle, Maureen K. "Problems with Conservation Tillage." *Journal of Soil and Water Conservation* 38 (1983): 201–6.

Hobbs, J. A., and P. L. Brown. *Nitrogen and Organic Carbon Changes in Cultivated Western Kansas Soils*. Kansas Agricultural Experiment Station Bulletin 89. Manhattan, Kans.: Kansas State University, 1957.

Hoffman, Carol A. "Ecological Risks of Genetic Engineering of Crop Plants." *BioScience* 40 (1990): 434–37.

Holling, C. S. "Resilience and Stability of Ecological Systems." *Annual Review of Ecology and Systematics* 4 (1973): 1–23.

Horn, David J. "Effect of Weedy Backgrounds on Colonization of Collards by Green Peach Aphid, *Myzus persicae*, and Its Major Predators." *Environmental Entomology* 10 (1981): 285–89.

Horsfall, James G. et al. *Genetic Vulnerability of Major Crops*. Washington, D.C.: National Academy of Sciences, 1972.

Horvitz, Carol C., and Douglas W. Schemske. "Demographic Cost of Reproduction in a Neotropical Herb: An Experimental Field Study." *Ecology* 69 (1988): 1741–45.

House, Garfield J., and Maria del Rosario Alzugaray. "Influence of Cover Cropping and No-tillage Practices on Community Composition of Soil Arthropods in a North Carolina Agroecosystem." *Environmental Entomology* 18 (1989): 302–7.

House, Garfield J., and Robert W. Parmelee. "Comparison of Soil Arthropods and Earthworms from Conventional and No-tillage Agroecosystems." *Soil and Tillage Research* 5 (1985): 351–60.

House, Garfield J., and Benjamin R. Stinner. "Agriculture and Ecosystems." In *Global Perspectives on Agroecology and Sustainable Agricultural Systems*, edited by Patricia Allen and Debra Van Dusen, 345–54. Santa Cruz: University of California, 1988.

Houston, P. "Waiting for Recovery Is Like Watching Corn Grow." *Business Week*, 12 January 1987, 82–83.

Howard, Sir Albert. *An Agricultural Testament*. New York: Oxford University Press, 1943.

Howe, C. W., and W. A. Ahrens. "Water Resources of the Upper Colorado River Basin: Problems and Policy Alternatives." In *Water and Arid Lands of the Western United States*, edited by Mohamed T. El-Ashry and Diana C. Gibbons, 169–232. World Resources Institute. New York: Cambridge University Press, 1988.

Hubbard, R. K., and J. M. Sheridan. "Nitrate Movement to Groundwater in the Southeastern Coastal Plain." *Journal of Soil and Water Conservation* 44 (1989): 20–27.

Hunter, A. F., and L. W. Aarssen. "Plants Helping Plants." *BioScience* 38 (1988): 34–40.

Iltis, Hugh H. "Tropical Forests: What Will Be Their Fate?" *Environment* 25(10) (1983): 55–60.

Inghe, Ola, and Carl Olaf Tamm. "Survival and Flowering of Perennial Herbs. IV. The Behaviour of *Hepatica nobilis* and *Sanicula europaea* on Permanent Plots During 1943–1981." *Oikos* 45 (1988): 400–20.

———. "Survival and Flowering of Perennial Herbs. V. Patterns of Flowering." *Oikos* 51 (1988): 203–19.

Ingram, G. B., and J. T. Williams. "In Situ Conservation of Wild Relatives of

Crops." In *Crop Genetic Resources: Conservation and Evaluation*, edited by J. H. W. Holden and J. T. Williams, 163–79. London: George Allen and Unwin, 1984.

"Insurance Company Farm Ownership Estimated at 5 Million Acres." *Land Stewardship Letter* 6(3) (1988): 3.

"Iowa Low-input Projects Demonstrate Effective Farm Savings." *Alternative Agriculture News* 7(5) (1989): 2.

Issar, A. "The Role of Non-replenishable Aquifers in Development Projects in Arid Regions." In *Settling the Desert*, edited by L. Berkofsky, D. Faiman, and J. Gales, 117–28. New York: Gordon and Breach Science Publishers, 1981.

Jackson, Laura L. "Life History Consequences of Greater Seed Production in a Perennial Grass, *Tripsacum dactyloides:* A Comparison of High and Low Seed-yielding Genotypes." Ph.D. diss., Cornell University, 1990.

Jackson, Wes. *New Roots for Agriculture*. New ed. Lincoln, Neb.: University of Nebraska Press, 1985.

———. *Altars of Unhewn Stone*. San Francisco: North Point Press, 1987.

———. "Agriculture with Nature as Analogy." In *Sustainable Agriculture in Temperate Zones*, edited by Charles A. Francis, Cornelia Butler Flora, and Larry D. King, 381–422. New York: John Wiley and Sons, 1990.

Jackson, Wes, and Jon Piper. "The Necessary Marriage Between Ecology and Agriculture." *Ecology* 70 (1989): 1591–93.

Jarosz, A. M., and M. Levy. "Effects of Habitat and Population Structure on Powdery Mildew Epidemics in Experimental *Phlox* Populations." *Phytopathology* 78 (1988): 358–62.

Jenkins, Robert E. "Long-term Conservation and Preserve Complexes." *Nature Conservancy* 39 (January–February 1989): 5–7.

Jennersten, O., S. G. Nilsson, and U. Wastljung. "Local Plant Populations as Ecological Islands: The Infection of *Viscaria vulgaris* by the Fungus *Ustilago violacea*." *Oikos* 41 (1983): 391–95.

Jin, Yisheng, Mingzhong Lu, and Chengyong Sun. "Preliminary Evaluation of Energy Consumption in China's Agriculture—Analysis of Two Specific Cases." *Energy in Agriculture* 1 (1983): 289–301.

Jordan, John Patrick, Paul F. O'Connell, and Roland R. Robinson. "Historical Evolution of the State Agricultural Experiment Station System." In *New Directions for Agriculture and Agricultural Research*, edited by Kenneth A. Dahlberg, 147–62. Totowa, N.J.: Rowman and Wastljung, 1983.

Kahn, Ephraim. "Pesticide Related Illness in California Farm Workers." *Journal of Occupational Medicine* 18 (1976): 693–96.

Kaiser, C. J. "C-4 Grass with Legumes: An Energy Efficient Biomass System." In *Proceedings of the XVIth International Grassland Congress*, 457–58. Nice, France, 1989.

Kannenberg, L. W. "Utilization of Genetic Diversity in Crop Breeding." In

Plant Genetic Resources: A Conservation Imperative, edited by Christopher W. Yeatman, David Kafton, and Garrison Wilkes, 93–109. Boulder: Westview Press, 1984.

Keeney, Dennis R. "Toward a Sustainable Agriculture: Need for Clarification of Concepts and Terminology." *American Journal of Alternative Agriculture* 4 (1989): 101–5.

Keller, Evelyn Fox. *Reflections on Gender and Science.* New Haven, Conn.: Yale University Press, 1985.

King, Juliana. "For Farm Finances: Promising Signs of a Cooling Crisis." *Farmline* (April 1987): 6–7, 12.

Knapp, Alan K., and Timothy R. Seastedt. "Detritus Accumulation Limits Productivity of Tallgrass Prairie." *BioScience* 36 (1986): 662–68.

Knudsen, Thomas J. "Experts Say Chemicals Peril Farm Belt's Water." *New York Times,* 6 September 1986.

Komarov, V. L., ed. *Flora of the U.S.S.R.* Vol. 3, *Gramineae.* Leningrad: Botanical Institute of the Academy of Sciences of the U.S.S.R., 1934. (Translated by the Office of Technical Services, U.S. Department of Commerce, Washington, D.C., 1963.)

Korsching, Peter F., and Paul Lasley. "Problems in Identifying Rural Unemployment." *Rural Sociologist* 6 (1986): 171–80.

Kovda, Victor A. "Arid Land Irrigation and Soil Fertility: Problems of Salinity, Alkalinity, Compaction." In *Arid Land Irrigation in Developing Countries,* edited by E. B. Worthington, 211–35. Oxford: Pergamon Press, 1977.

Kowalski, R. "Organic Farming—A Sound Basis for Pest Management?" In *Towards a Sustainable Agriculture,* edited by J. M. Besson and H. Vogtmann, IFOAM International Conference. Sissach, Switzerland: Verlag Wirz Auarau, 1978.

Kowalski, R., and P. E. Visser. "Nitrogen in a Crop-pest Interaction: Cereal Aphids 1979." In *Nitrogen as an Ecological Factor,* edited by J. A. Lee, S. McNeill, and I. H. Rorison, 283–300. The 22nd Symposium of the British Ecological Society. Oxford: Blackwell Scientific Publications, 1983.

Kraus, Tamara. "Evaluation of Overwintering Capabilities in Sorghum Crosses: *Sorghum bicolor* × *Sorghum halepense.*" *The Land Institute Research Report* 7 (1990): 31–33.

Kroll, T. K., L. D. Moore, and G. H. Lacy. "Interplanting Susceptible and Resistant Radish Cultivars Reduces Colonization by *Plasmodiophora brassicae.*" *HortScience* 19 (1984): 403–4.

Kulakow, Peter A., Laura L. Benson, and Jake G. Vail. "Prospects for Domesticating Illinois Bundleflower." In *Advances in New Crops,* edited by Jules Janick and James E. Simon, 168–71. Portland, Oreg.: Timber Press, 1990.

Lacy, William B., Laura R. Lacy, and Lawrence Busch. "Agricultural Biotechnology Research: Practices, Consequences, and Policy Recommendations." *Agriculture and Human Values* 5 (1988): 3–14.

Land Stewardship Project. "The 1990 Farm Bill: Environment Wins, Family Farmers Lose." *Land Stewardship Update* 1(3) (Winter 1990).

Lappé, Frances Moore, and Joseph Collins. *World Hunger: Twelve Myths.* New York: Food First and Grove Press, 1986.

Larson, William E., F. J. Pierce, and R. H. Dowdy. "The Threat of Soil Erosion to Long-term Crop Production." *Science* 219 (1983): 458–65.

Lecos, Chris. "Pesticides and Food: Public Worry No. 1." *FDA Consumer* 18(6) (1984): 12–15.

Leistritz, F. Larry, and Brenda L. Ekstrom. "The Financial Characteristics of Production Units and Producers Experiencing Financial Stress." In *The Farm Financial Crisis,* edited by Steve H. Murdock and F. Larry Leistritz, 73–95. Boulder: Westview Press, 1988.

Leistritz, F. Larry, Brenda L. Ekstrom, Harvey G. Vreugdenhil, and Janet Wanzek. "Producer Reactions and Adaptations." In *The Farm Financial Crisis,* edited by Steve H. Murdock and F. Larry Leistritz, 97–111. Boulder: Westview Press, 1988.

Leistritz, F. Larry, and Steve H. Murdock. "Financial Characteristics of Farms and of Farm Financial Markets and Policies in the United States." In *The Farm Financial Crisis,* edited by Steve H. Murdock and F. Larry Leistritz, 13–28. Boulder: Westview Press, 1988.

Leonard, K. J. "Factors Affecting Rates of Stem Rust Increase in Mixed Plantings of Susceptible and Resistant Oat Varieties." *Phytopathology* 59 (1969): 1845–50.

Leonard, L. T. "Lack of Nodule Formation in a Subfamily of the Leguminosae." *Soil Science* 20 (1925): 165–67.

Lewontin, Richard C., and Jean-Pierre Barlan. "The Political Economy of Agricultural Research: The Case of Hybrid Corn." In *Agroecology,* edited by C. Ronald Carroll, John H. Vandermeer, and Peter M. Rosset, 613–28. New York: McGraw-Hill, 1990.

Liebman, Matthew. "Ecological Suppression of Weeds in Intercropping Systems: A Review." In *Weed Management in Agroecosystems: Ecological Approaches,* edited by Miguel A. Altieri and Matthew Liebman, 198–212. Boca Raton, Fla.: CRC Press, 1988.

Lindemann, W. C., G. W. Randall, and George E. Ham. "Tillage Effects on Soybean Nodulation, $N_2(C_2H_4)$ Fixation, and Seed Yield." *Agronomy Journal* 74 (1982): 1067–70.

Little, Charles E. *Green Fields Forever.* Washington,D.C.: Island Press, 1987.

Lockeretz, William. "Establishing the Proper Role for On-farm Research." *American Journal of Alternative Agriculture* 2 (1987): 132–36.

Lofton, S. M. "Nitrogen Fixation in Selected Prairie Legumes as Related to Succession." Ph.D. diss., Oklahoma State University, 1976.

Lovins, Amory B., L. Hunter Lovins, and Marty Bender. "Energy and Agriculture." In *Meeting the Expectations of the Land,* edited by Wes Jackson,

Wendell Berry, and Bruce Colman, 68–86. San Francisco: North Point Press, 1984.

Luck, Robert F. "Evaluation of Natural Enemies for Biological Control: A Behavioral Approach." *Trends in Ecology and Evolution* 5 (1990): 196–99.

Luck, Robert F., B. Merle Shepard, and Peter E. Kenmore. "Experimental Methods for Evaluating Arthropod Natural Enemies." *Annual Review of Entomology* 33 (1988): 367–91.

Maass, J. M., C. F. Jordan, and J. Sarukhan. "Soil Erosion and Nutrient Losses in Seasonal Tropical Agroecosystems under Various Management Techniques." *Journal of Applied Ecology* 25 (1988): 595–607.

Mack, Richard N., and John N. Thompson. "Evolution in Steppe with Few Large, Hooved Mammals." *American Naturalist* 119 (1982): 218–21.

MacMahon, James A. "Warm Deserts." In *North American Terrestrial Vegetation*, edited by Michael G. Barbour and William Dwight Billings, 231–64. Cambridge: Cambridge University Press, 1988.

MacRae, Rod J., Stuart B. Hill, John Henning, and Guy R. Mehuys. "Agricultural Science and Sustainable Agriculture: A Review of the Existing Scientific Barriers to Sustainable Food Production and Potential Solutions." *Biological Agriculture and Horticulture* 6 (1989): 173–219.

Madden, Patrick. "Beyond Conventional Economics—An Examination of the Values Implicit in the Neoclassical Economic Paradigm as Applied to the Evaluation of Agricultural Research." In *New Directions for Agriculture and Agricultural Research*, edited by Kenneth A. Dahlberg, 221–58. Totowa, N.J.: Rowman and Allanheld, 1986.

McBride, Bob. "Broken Heartland." *The Nation*, 8 February 1986, 132–33.

McCracken, R. J., J. Lee, R. W. Arnold, and D. E. McCormack. "An Appraisal of Soil Resources in the USA." In *Soil Erosion and Crop Productivity*, edited by R. F. Follett and B. A. Stewart, 55–67. Madison, Wisc.: ASA-CSSA-SSSA, 1985.

McGovern, George. *Agricultural Thought in the Twentieth Century.* Indianapolis: Bobbs-Merrill, 1967.

McKenry, Michael V. "Potential Problems with Soil Applied Chemicals." In *Pesticides in Soil and Groundwater*, edited by Harold G. Alford and Mary P. Ferguson, 15–21. Berkeley: Division of Agricultural Sciences, University of California, 1983.

McNamara, Ken. "Research Update: The Big Difference Was Cost." *New Farm* 11 (January 1989): 17–20.

McWilliams, Lindsey. "A Bumper Crop Yields Growing Problems." *Environment* 26(4) (1984): 25–34.

Merchant, Carolyn. *Ecological Revolutions.* Chapel Hill, N.C.: University of North Carolina Press, 1989.

Mervis, J. "Panel Weighs Overhaul of NSF's Grant System." *The Scientist* 5 (1991): 1, 6–7, 12.

Meyers, C. F., J. Meek, S. Tuller, and A. Weinberg. "Nonpoint Sources of Water Pollution." *Journal of Soil and Water Conservation* 40 (1985): 19–26.

Miller, Fred P., Wayne D. Rasmussen, and L. Donald Meyer. "Historical Perspective of Soil Erosion in the United States." In *Soil Erosion and Crop Productivity*, edited by R. F. Follett and B. A. Stewart, 23–48. Madison, Wisc.: ASA-CSSA-SSSA, 1985.

Mkpadi, M. C. "Energy for Agriculture in Nigeria—A Future Problem." *Energy in Agriculture* 6 (1987): 273–80.

Moore, James W. *Balancing the Needs of Water Use.* New York: Springer-Verlag, 1989.

Mundt, Christopher C., and J. A. Browning. "Development of Crown Rust Epidemics in Genetically Diverse Oat Populations: Effect of Genotype Unit Area." *Phytopathology* 75 (1985): 607–10.

Mundt, Christopher C., and K. J. Leonard. "Effect of Host Genotype Unit Area on Epidemic Development of Crown Rust Following Focal and General Inoculations of Mixtures of Immune and Susceptible Host Plants." *Phytopathology* 75 (1985): 1141–45.

———. "Effect of Host Genotype Unit Area on Development of Focal Epidemics of Bean Rust and Common Maize Rust in Mixtures of Resistant and Susceptible Plants." *Phytopathology* 76 (1986): 895–900.

Murdock, Steve H., Rita R. Hamm, Lloyd B. Potter, and Don E. Albrecht. "Demographic Characteristics of Rural Residents in Financial Distress and Social and Community Impacts of the Farm Crisis." In *The Farm Financial Crisis*, edited by Steve H. Murdock and F. Larry Leistritz, 113–40. Boulder: Westview Press, 1988.

Murdock, Steve H., and F. Larry Leistritz. "Policy Alternatives and Research Agenda." In *The Farm Financial Crisis*, edited by S. H. Murdock and F. L. Leistritz, 169–84. Boulder: Westview Press, 1988.

Murphy, Robert, and Clair Harvey. "Residues and Metabolites of Selected Persistent Halogenated Hydrocarbons in Blood Specimens from a General Population Survey." *Environmental Health Perspectives* 60 (1985): 115–20.

Muto, Paul J. "Evaluation of Illinois Bundleflower and Eastern Gamagrass Germplasm in Monocultures and Bicultures." *The Land Institute Research Report* 7 (1990): 26–30.

Myers, N. "Saving Wild Genes." *International Wildlife* 13(6) (1983): 24.

Nabhan, Gary Paul. *The Desert Smells Like Rain.* San Francisco: North Point Press, 1982.

———. *Gathering the Desert.* Tucson: University of Arizona Press, 1985.

National Research Council. *Alternative Agriculture.* Washington, D.C.: National Academy Press, 1989.

Nations, James D., and Daniel I. Komer. "Rainforests and the Hamburger Society." *Environment* 25(3) (1983): 12–20.

Nauta, R. S. "Agricultural Production in the Netherlands by Natural Supply of Nitrogen." *Biological Agriculture and Horticulture* 4 (1987): 181–201.

Ness, Richard. "A Paradigm Shift for Extension: On-farm Research in Sustainable Farming Practices." *Land Stewardship Letter* 7 (1981): 12–14.

"New Moves to Limit EDB in Food." *Science News* 125 (1984): 89.

Nowak, Peter J. "The Costs of Excessive Soil Erosion."*Journal of Soil and Water Conservation* 43 (1988): 307–10.

O'Connell, Paul. F. "National Agenda Setting and the State Agricultural Experiment Stations." In *Research Perspective*, edited by Experiment Station Committee on Organization and Policy. Proceedings of Symposium on the Research Agenda for the State Agricultural Experiment Stations. College Station, Tex.: Texas Agricultural Experiment Station and Texas A & M University, 1985.

O'Donnell, M. S., and T. H. Coaker. "Potential of Intracrop Diversity for the Control of Brassica Pests." *Proceedings of the 8th British Insecticide and Fungicide Conference* 1 (1975): 101–5.

Odum, Eugene P. "The Strategy of Ecosystem Development." *Science* 164 (1969): 262–70.

Odum, Howard T. "Trophic Structure and Productivity of Silver Springs, Florida." *Ecological Monographs* 27 (1957): 55–112.

———. "Self-organization, Transformity, and Information." *Science* 242 (1988): 1132–39.

Osterman, Douglas A., and Theresa L. Hicks. "Highly Erodible Land: Farmer Perceptions Versus Actual Measurements." *Journal of Soil and Water Conservation* 43 (1988): 177–81.

Papendick, Robert I., P. A. Sanchez, and G. B. Triplett, eds. *Multiple Cropping*. ASA Special Publication No. 27. Madison, Wisc.: ASA-CSSA-SSSA, 1976.

Papendick, Robert I., D. L. Young, D. K. McCool, and H. A. Krauss. "Regional Effects of Soil Erosion on Crop Productivity—The Palouse Area of the Pacific Northwest." In *Soil Erosion and Crop Productivity*, edited by R. F. Follett and B. A. Stewart. Madison, Wisc.: ASA-CSSA-SSSA, 1985.

Parker, Matthew A. "Genetic Uniformity and Disease Resistance in a Clonal Plant." *American Naturalist* 132 (1988): 538–49.

Parmelee, Robert W., and D. A. Crossley, Jr. "Earthworm Production and Role in the Nitrogen Cycle of a No-tillage Agroecosystem on the Georgia Piedmont." *Pedobiologia* 32 (1988): 353–61.

Parrish, Judith A. D., and Fakhri A. Bazzaz. "Underground Niche Separation in Successional Plants." *Ecology* 57 (1976): 1281–88.

———. "Competitive Interactions in Plant Communities of Different Successional Ages." *Ecology* 63 (1982): 314–20.

Pasternak, D., J. A. Aronson, J. Ben-Dov, M. Forti, S. Mendlinger, A. Nerd, and D. Sitton. "Development of New Arid Zone Crops for the Negev Desert of Israel." *Journal of Arid Environments* 11 (1986): 37–59.

Paustian, K. O. Andren, M. Clarholm, A.-C. Hansson, G. Johansson, J. Lagerhof, T. Lindberg, R. Pettersson, and B. Sohlenius. "Carbon and Nitrogen Budgets of Four Agro-ecosystems with Annual and Perennial Crops, with and without N. Fertilization." *Journal of Applied Ecology* 27 (1990): 60–84.

Perrin, Noel. "The High Price of Agribusiness." *Land Stewardship Letter* 9 (Winter 1991): 6–7.

Perry, D. A., M. P. Amaranthus, J. G. Borchers, S. L. Borchers, and R. E. Brainerd. "Bootstrapping in Ecosystems." *BioScience* 39 (1989): 230–37.

"Pesticide Is Banned, But Congress Asks Why It Took So Long." *Science News* 124 (1983): 229.

Peterson, C. "Farm Water Poisons Wildlife." *Washington Post*, 10 March 1985.

Petrova, K. A. "Hybridization between Wheat and *Elymus*." In *Wide Hybridization in Plants*, edited by N. V. Tsitsin, 226–37. Academy of Sciences of the U.S.S.R. Moscow: All-Union Academy of Agricultural Sciences, 1960. (Published for the National Science Foundation, Washington, D.C., by the Israeli Program for Scientific Translations, Jerusalem, 1962.)

"Pillsbury Wins Insurance Ruling." *New York Times*, 9 December 1988.

Pimentel, David. *Ecological Effects of Pesticides on Non-target Species.* Washington, D.C.: U. S. Government Printing Office, 1971.

———. "The Ecological Basis of Insect Pest, Pathogen and Weed Problems." In *Origins of Pest, Parasite, Disease and Weed Problems*, edited by J. M. Cherrett and G. R. Sagar, 3–31. The Eighteenth Symposium of the British Ecological Society. Oxford: Blackwell Scientific, 1977.

———. "Introduction." In *Handbook of Energy Utilization in Agriculture*, edited by D. Pimentel, 3–6. Boca Raton, Fla.: CRC Press, 1980.

———. "Energy Flow in Agroecosystems." In *Agricultural Ecosystems*, edited by R. Lowrance, B. R. Stinner, and G. J. House, 121–32. New York: John Wiley and Sons, 1984.

Pimentel, David, David Andow, David Gallahan, Ilse Schreiner, Todd E. Thompson, Rada Dyson-Hudson, Stuart Neil Jacobson, Mary Ann Irish, Susan F. Kroop, Anne M. Moss, Michael D. Shepard, and Billy G. Vinzant. "Pesticides: Environmental and Social Costs." In *Pest Control: Cultural and Environmental Aspects*, edited by David Pimentel and J. H. Perkins, 99–158. Boulder: Westview Press, 1980.

Pimentel, David, and Michael Burgess. "Energy Inputs in Corn Production." In *Handbook of Energy Utilization in Agriculture*, edited by D. Pimentel, 67–84. Boca Raton, Fla.: CRC Press, 1980.

———. "Effects of Single Versus Combinations of Insecticides on the Development of Resistance." *Environmental Entomology* 14 (1985): 582–89.

Pimentel, David, and Wen Dazhong. "Technological Changes in Energy Use in U.S. Agricultural Production." In *Agroecology*, edited by C. Ronald Carroll, John H. Vandermeer, and Peter M. Rosset, 147–64. New York: McGraw-Hill, 1990.

Pimentel, David, Sarah Fast, Wei Liang Chao, Ellen Stuart, Joanne Dintzis,

Gail Einbender, William Schlappi, David Andow, and Kathryn Broderick. "Water Resources in Food and Energy." *BioScience* 32 (1982): 861–67.

Pimentel, David, Carol Glenister, Sarah Fast, and David Gallahan. "An Environmental Risk Assessment of Biological and Cultural Controls for Organic Agriculture." In *Environmentally Sound Agriculture,* edited by W. Lockeretz, 73–90. New York: Praeger, 1983.

Pimentel, David, M. S. Hunter, J. A. LaGro, R. A. Efroymson, J. C. Landers, F. T. Mervis, C. A. McCarthy, and A. E. Boyd. "Benefits and Risks of Genetic Engineering in Agriculture." *BioScience* 39 (1989): 606–14.

Pimentel, David, John Krummel, David Gallahan, Judy Hough, Alfred Merrill, Ilse Schreiner, Pat Vittum, Fred Koziol, Ephraim Back, Doreen Yen, and Sandy Fiance. "A Cost-benefit Analysis of Pesticide Use in U.S. Food Production." In *Pesticides: Contemporary Roles in Agriculture, Health, and Environment,* edited by T. J. Sheets and D. Pimentel, 97–149. Clifton, N.J.: Humana Press, 1979.

Piper, Jon K. "Germination and Growth of Bird-dispersed Plants: Effects of Seed Size and Light on Seedling Vigor and Biomass Allocation." *American Journal of Botany* 73 (1986): 959–65.

Piper, Jon K., and Mark C. Gernes. "Vegetation Dynamics of Three Tallgrass Prairie Sites." In *Proceedings of the Eleventh North American Prairie Conference,* edited by T. Bragg and J. Stubbendieck, 9–14. Lincoln: University of Nebraska Press, 1990.

Piper, Jon, James Henson, Mary Bruns, and Marty Bender. "Seed Yield and Quality Comparison of Herbaceous Perennials and Annual Crops." In *Global Perspectives on Agroecology and Sustainable Agricultural Systems,* edited by P. Allen and D. Van Dusen, 715–19. Santa Cruz: University of California, 1988.

Piper, Jon K., and Douglas Towne. "Multiple Year Patterns of Seed Yield in Five Herbaceous Perennials." *The Land Institute Research Report* 5 (1988): 14–18.

Pitelka, Louis F., Sandra B. Hansen, and Jeffry W. Ashmun. "Population Biology of *Clintonia borealis.* I. Ramet and Patch Dynamics." *Journal of Ecology* 73 (1985): 169–83.

Plucknett, D. L., and N. J. H. Smith. "Historical Perspectives on Multiple Cropping." In *Multiple Cropping Systems,* edited by Charles A. Francis, 20–39. New York: Macmillan, 1986.

Plucknett, Donald L., Nigel J. H. Smith, J. T. Williams, and N. M. Anishetty. "Crop Germplasm Conservation and Developing Countries." *Science* 220 (1983) 163–69.

Postel, Sandra. "Managing Fresh Water Supplies." In *The State of the World 1985,* 42–72. Washington, D.C.: Worldwatch Institute, 1985.

———. *Conserving Water: The Untapped Alternative.* Washington, D.C.: Worldwatch Institute, 1985.

Price, Peter W. "An Overview of Organismal Interactions in Ecosystems in Evolutionary and Ecological Time." *Agriculture, Ecosystems and Environment* 24 (1988): 369–77.

Rabinowitz, Deborah, Jody K. Rapp, Victoria L. Sork, Beverly J. Rathcke, Gary A. Reese, and Jan C. Weaver. "Phenological Properties of Wind- and Insect-pollinated Prairie Plants." *Ecology* 62 (1981): 49–56.

Raloff, J. "EPA to Limit, Then Ban EDB in Citrus." *Science News* 125 (1984): 151.

Rao, M. R., and R. W. Willey. "Evaluation of Yield Stability in Intercropping: Studies on Sorghum/Pigeonpea." *Experimental Agriculture* 16 (1980): 105–16.

Rapport, David J. "What Constitutes Ecosystem Health?" *Perspectives in Biology and Medicine* 33 (1989): 120–32.

Rasmussen, Wayne D., ed. *Readings in the History of American Agriculture.* Urbana: University of Illinois Press, 1960.

Reekie, E. G., and Fakhri A. Bazzaz. "Reproductive Effort in Plants. 3. Effect of Reproduction on Vegetative Activity." *American Naturalist* 129 (1987): 907–19.

Rees, William E. "The Ecology of Sustainable Development." *The Ecologist* 20 (1990): 18–23.

Reimann-Philipp, R. "Perennial Spring Rye as a Crop Alternative." *Journal of Agronomy and Crop Science* 157 (1986): 281–85.

Rhoades, David F., and Rex G. Cates. "Toward a General Theory of Plant Antiherbivore Chemistry." *Recent Advances in Phytochemistry* 10 (1976): 168–213.

Rice, Elroy L. *Allelopathy.* 2d ed. Orlando: Academic Press, 1984.

Risch, Stephen J. "The Population Dynamics of Several Herbivorous Beetles in a Tropical Agroecosystem: The Effect of Intercropping Corn, Beans and Squash in Costa Rica." *Journal of Applied Ecology* 17 (1980): 593–612.

———. "Insect Herbivore Abundance in Tropical Monocultures and Polycultures: An Experimental Test of Two Hypotheses." *Ecology* 62 (1981): 1325–40.

Risch, Stephen J., David Andow, and Miguel A. Altieri. "Agroecosystem Diversity and Pest Control: Data, Tentative Conclusions, and New Research Directions." *Environmental Entomology* 12 (1983): 625–29.

Risser, P. G., E. C. Birney, H. D. Blocker, S. W. May, W. J. Parton, and J. A. Wiens. *The True Prairie Ecosystem.* United States/International Biological Program Synthesis Series. Vol. 16. Stroudsburg, Pa.: Hutchinson Ross Publishing Company, 1981.

Rogers, Susan Carol. "Mixing Paradigms on Mixed Farming: Anthropological and Economic Views of Specialization in Illinois Agriculture." In *Farm Work and Fieldwork,* edited by Michael Chibnik, 58–89. Ithaca, N.Y.: Cornell University Press, 1987.

Romig, Doug. "Sustainable Agriculture: Avoiding the Definition Trap." *The Land Report* 39 (Fall 1990): 14–15.

Root, Richard B. "Organization of a Plant-arthropod Association in Simple and Diverse Habitats: The Fauna of Collards (*Brassica oleracea*)." *Ecological Monographs* 43 (1973): 95–124.

Rowe, J. Stan. "The Level-of-integration Concept and Ecology." *Ecology* 42 (1961): 420–27.

———. "Land and People: An Ecological Perspective." *The Land Report* 33 (Summer 1988): 10–15.

Runkle, J. R. "Patterns of Disturbance in Some Old-growth Mesic Forests in Eastern North America." *Ecology* 63 (1982): 1533–46.

Rutger, J. N., and W. R. Grant. "Energy Use in Rice Production." In *Handbook of Energy Utilization in Agriculture*, edited by D. Pimentel, 93–98. Boca Raton, Fla.: CRC Press, 1980.

Ruttan, Vernon W. "Agricultural Research Policy Concerns." In *Agricultural Science Policy in Transition*, edited by V. J. Rhodes, 103–22. Bethesda, Md.: Agricultural Research Institute, 1986.

Sagoff, Mark. "Biotechnology and the Environment: What Is at Risk?" *Agriculture and Human Values* 5 (1988): 26–35.

Sahtouris, Elisabet. *Gaia: The Human Journey from Chaos to Cosmos*. New York: Pocket Books, 1989.

Salamon, Sonya. "Ethnic Determinants of Farm Community Character." In *Farm Work and Fieldwork*, edited by Michael Chibnik, 167–88. Ithaca, N.Y.: Cornell University Press, 1987.

Salick, Jan, and Laura C. Merrick. "Use and Maintenance of Genetic Resources: Crops and Their Wild Relatives." In *Agroecology*, edited by C. Ronald Carroll, John H. Vandermeer, and Peter M. Rosset, 517–48. New York: McGraw-Hill, 1990.

Satchell, J. E. "Earthworm Biology and Soil Fertility." *Soils and Fertilizers* 21 (1958): 209–19.

Schertz, David L. "Conservation Tillage: An Analysis of Acreage Projections in the United States." *Journal of Soil and Water Conservation* 43 (1988): 256–58.

Schmidt, W. E. "Iowans Struggle with Growing Water Pollution." *New York Times*, 22 November 1987.

Schoenau, J. J., J. W. B. Stewart, and J. R. Bettany. "Forms of Cycling of Phosphorus in Prairie and Boreal Forest Soils." *Biogeochemistry* 8 (1989): 223–37.

Schwab, A. P., C. E. Owensby, and S. Kulyingyong. "Changes in Soil Chemical Properties Due to 40 Years of Cultivation." *Soil Science* 149 (1990): 35–43.

Seastedt, Timothy R., Rosemary A. Ramundo, and David C. Hayes. "Maximization of Densities of Soil Animals by Foliage Herbivory: Empirical Evidence, Graphical and Conceptual Models." *Oikos* 51 (1988): 243–48.

Segal, A., J. Manisterski, G. Fishbeck, and I. Wahl. "How Plant Populations Defend Themselves in Natural Ecosystems." In *Plant Disease: An Advanced Treatise. Vol. 5. How Plants Defend Themselves,* edited by James G. Horsfall and Ellis B. Cowling, 75–102. New York: Academic Press, 1980.

Shaner, G. "Effect of Environment on Fungal Leaf Blights of Small Grains." *Annual Review of Phytopathology* 19 (1981): 273–96.

Sheehy, J. E. "How Much Dinitrogen Fixation Is Required in Grazed Grassland?" *Annals of Botany* 64 (1989): 159–61.

Shepherd, James F. "Soil Conservation in the Pacific Northwest Wheat-Producing Areas: Conservation in a Hilly Terrain." In *Agricultural History,* edited by D. Helms and Susan L. Flader, 127–43. Berkeley: University of California Press, 1985.

Silva, J. Sousa. "The Contradictions of the Biorevolution for the Development of Agriculture in the Third World: Biotechnology and Capitalist Interest." *Agriculture and Human Values* 5 (1988): 61–70.

Simonian, Lane. "Pesticide Use in Mexico: Decades of Abuse." *The Ecologist* 18 (1988): 82–87.

Singh, Bijay, and G. S. Sekhon. "Nitrate Pollution of Groundwater from Farm Use of Nitrogen Fertilizers—A Review." *Agriculture and Environment* 4 (1978–1979): 207–25.

Smartt, J. "Gene Pools in Grain Legumes." *Economic Botany* 38 (1984): 24–35.

Smith, Judith G. "Influence of Crop Background on Natural Enemies of Aphids on Brussels Sprouts." *Annals of Applied Biology* 83 (1976): 15–29.

Smith, J. Russell. *Tree Crops: A Permanent Agriculture.* New York: The Devin-Adair Company, 1953.

Spedding, C. R. W. "Agricultural Systems of the Future." In *Energy Management and Agriculture,* edited by D. W Robinson and R. C. Mollan, 419–30. Proceedings of the First International Summer School in Agriculture, held in cooperation with the W. K. Kellogg Foundation. Dublin: Royal Dublin Society, 1982.

"Stewardship Committee Surveys Insurance Companies' Farm Ownership Policies." *Land Stewardship Letter* (Spring 1988): 5.

Stokes, B. "Water Shortages: The Next Energy Crisis." *The Futurist* 17 (April 1983): 38–47.

Stokes, C. Shannon, and Kathy D. Brace. "Agricultural Chemical Use and Cancer Mortality in Selected Rural Counties in the U.S.A." *Journal of Rural Studies* 4 (1988): 239–47.

Stout, B. A., with C. A. Myers, A. Hurand, and L. W. Faidley. *Energy for World Agriculture.* FAO Agriculture Series No. 7. Rome: Food and Agriculture Organization of the United Nations, 1979.

Strange, Marty. *Family Farming: A New Economic Vision*. Lincoln: University of Nebraska Press, 1988.

Sun, Marjorie. "EDB Contamination Kindles Federal Action." *Science* 223 (1984): 464–66.

——. "Ground Water Ills: Many Diagnoses, Few Remedies." *Science* 232 (1986): 1490–93.

Taff, Steven J., and C. Ford Runge. "Wanted: A Leaner and Meaner CRP." *Choices* (First Quarter 1988): 16–18.

Tahvanainen, J. O., and Richard B. Root. "The Influence of Vegetational Diversity on the Population Ecology of a Specialized Herbivore, *Phyllotreta cruciferae* (Coleoptera: Chrysomelidae)." *Oecologia* 10 (1972): 321–46.

Tawczynski, Dan. "Got a Problem? Ask a Farmer!" *New Farm* 12 (July–August 1990): 48.

Teeri, James A., and L. G. Stowe. "Climatic Patterns and the Distribution of C_4 Grasses in North America." *Oecologia* 23 (1976): 1–12.

"The Persistence of EDB in Soils." *Environment* 29 (10) (1987): 23–24.

Thompson, Richard C. "The Search for Pesticide Residues." *FDA Consumer* 18 (6) (1984): 6–10.

Thompson, T., D. Zimmerman, and C. Rogers. "Wild *Helianthus* as a Genetic Resource." *Field Crops Research* 4 (1981): 333–43.

Thornberg, A. *Grass and Legume Seed Production in Montana*. Montana State University Agricultural Experiment Station Bulletin 333. Bozeman: Montana State University, 1971.

Tilman, David. *Plant Strategies and the Dynamics and Structure of Plant Communities*. Princeton: Princeton University Press, 1988.

Trenbath, Brian R. "Resource Use by Intercrops." In *Multiple Cropping Systems*, edited by Charles A. Francis, 57–81. New York: Macmillan, 1986.

Tukahira, E. M., and T. H. Coaker. "Effect of Mixed Cropping on Some Insect Pests of Brassicas; Reduced *Brevicoryne brassicae* Infestations and Influences on Epigeal Predators and the Disturbance of Oviposition Behavior in *Delia brassicae*." *Entomologia Experimentalis et Applicata* 32 (1982): 129–40.

U.S. Department of Agriculture. *The Current Financial Condition of Farmers and Farm Lenders*. Agriculture Information Bulletin No. 490. Coordinated by D. H. Harrington and J. M. Stam. Washington, D.C.: U.S. Department of Agriculture, Economic Research Service, 1985.

U.S. Department of Agriculture. *Agricultural Statistics*. Washington, D.C.: U.S. Department of Agriculture, for sale by Superintendent of Documents, U.S. Government Printing Office, 1986.

U.S. Department of Agriculture. *Agricultural Finance Outlook and Situation Report*. Publication AFO-26. Washington, D.C.: U.S. Department of Agriculture, Economic Research Service, March 1986.

U.S. Department of Agriculture. *Financial Characteristics of U.S. Farms, Janu-*

ary 1, 1987. Agriculture Information Bulletin No. 525. Washington, D.C.: U.S. Department of Agriculture, Economic Research Service, August 1987.

U.S. Department of Agriculture. "Farm Numbers Decline Slightly." Economic Research Service. *Farmline* (October 1987): 12.

U.S. Department of Agriculture. "Trying Times for the Input Industries." *Farmline* (December–January 1988): 10–11.

U.S. Department of Agriculture. "Farm Population Down to 2 Percent." Economic Research Service. *Farmline* (April 1988): 16–17.

U.S. Department of Agriculture. *Financial Characteristics of U. S. Farms, January 1, 1988.* Agriculture Information Bulletin No. 551. Washington, D.C.: U.S. Department of Agriculture, Economic Research Service, October 1988.

U.S. Department of Agriculture. *Financial Characteristics of U. S. Farms, January 1, 1989.* Agriculture Information Bulletin No. 579. Washington, D.C.: U.S. Department of Agriculture, Economic Research Service, December 1989.

U.S. Geological Survey. *Groundwater Regions of the United States.* Geological Survey Water Supply Paper 2242. Washington, D.C.: U.S. Department of the Interior, Geological Survey, for sale by Superintendent of Documents, U.S. Government Printing Office, 1984.

U.S. Water Resources Council. *The Nation's Water Resources 1975–2000. VII: Water Quantity, Quality, and Related Land Considerations.* Washington, D.C.: Water Resources Council, for sale by Superintendent of Documents, U.S. Government Printing Office, 1978.

Van Chantfort, Eric. "Agriculture and the Larger Economy: When Links Become Liabilities." *Farmline* (July 1984): 14–17.

———. "Financial Survey Shows Farmers on an Upswing before Drought." *Farmline* (December–January 1989): 4–9.

Van der Plank, J. E. *Plant Diseases: Epidemics and Control.* New York: Academic Press, 1963.

Vandermeer, John. *The Ecology of Intercropping.* Cambridge: Cambridge University Press, 1989.

van Emden, H. F., and G. F. Williams. "Insect Stability and Diversity of Agroecosystems." *Annual Review of Entomology* 19 (1974): 455–75.

van Kessel, Christopher, Paul W. Singleton, and Heinz J. Hoben. "Enhanced N-transfer from Soybean to Maize by Vesicular Arbuscular Mycorrhizal (VAM) Fungi." *Plant Physiology* 79 (1985): 562–63.

Van Sloten, D. H. "The Genetic Resources of Leafy Vegetables." In *Crop Genetic Resources: Conservation and Evaluation,* edited by J. H. W. Holden and J. T. Williams, 240–48. London: George Allen and Unwin, 1984.

Vira, Shiraz, and Harry Riehle. "Conservation in the Palouse: An Economic Dilemma." In *Economics, Ethics, Ecology: Roots of Productive Conservation,* edited by W. E. Jeske, 450–54. Ankeny, Iowa: Soil Conservation Society of America, 1981.

Wagoner, Peggy. *Perennial Grain Research at the Rodale Research Center. 1987 Summary.* Emmaus, Pa.: Rodale Press, 1988.

————. "Perennial Grain Development—Past Efforts and Potential for the Future." *CRC Critical Reviews in Plant Sciences* 9 (1990): 381–408.

Wall, W. "Iowa Bill to Curb Groundwater Pollution Could Prompt Similar Steps Elsewhere." *Wall Street Journal,* 13 March 1987.

Waters, W. F. "Another Look at Pesticides and Peasants." *The Rural Sociologist* 3 (1983): 259–64.

Weaver, John E. *Prairie Plants and Their Environment.* Lincoln: University of Nebraska Press, 1968.

Weaver, John E., and T. J. Fitzpatrick. "The Prairie." *Ecological Monographs* 4 (1934): 109–295.

Weltzien, H. C., and A. Trankner. "Species Mixtures in Cereals as a Concept for Nonchemical Disease and Pest Management." In *The Importance of Biological Agriculture in a World of Diminishing Resources,* edited by H. Vogtmann, E. Boehncke, and I. Fricke, 436–42. Witzenhausen, Germany: Verlagsgruppe Witzenhausen, 1986.

de Wet, J. M. J., and J. R. Harlan. "*Tripsacum* and the Origin of Maize." In *Maize Breeding and Genetics,* edited by David B. Walden, 129–41. New York: John Wiley and Sons, 1978.

Wheeler, William A. *Grassland Seeds.* New York: D. van Nostrand, 1957.

Whittaker, Robert H., and Simon A. Levin. "The Role of Mosaic Phenomena in Natural Communities." *Theoretical Population Biology* 12 (1977): 117–39.

Wilk, Valerie A. *The Occupational Health of Migrant and Seasonal Farmworkers in the United States.* Washington, D.C.: Farmworkers Justice Fund, Inc., 1986.

Williams, J. T. "A Decade of Crop Genetic Resources Research." In *Crop Genetic Resources: Conservation and Evaluation,* edited by J. H. W. Holden and J. T. Williams, 1–17. London, Boston: George Allen and Unwin, 1984.

Wilson, Edward O. "Outcry from a World of Wounds." *Discover* 6 (1985): 64–66.

————. "Threats to Biodiversity." *Scientific American* 261 (1989): 108–16.

Wittwer, Sylvan H. "Research and Technology Needs for the Twenty-First Century." In *Global Aspects of Food Production,* edited by M. S. Swaminathan and S. K. Sinha, 85–116. Oxford: Tycooly International, 1986.

Wolfe, M. S. "The Current Status and Prospects of Multiline Cultivars and Variety Mixtures for Resistance." *Annual Review of Phytopathology* 23 (1985): 251–73.

Wolman, M. G. "Soil Erosion and Crop Productivity: A Worldwide Perspective." In *Soil Erosion and Crop Productivity,* edited by R. F. Follett and B. A. Stewart, 9–21. Madison, Wisc.: ASA-CSSA-SSSA, 1985.

Woodmansee, R. G. "Factors Influencing Input and Output of Nitrogen in

Grasslands." In *Perspectives in Grassland Ecology,* edited by N. R. French, 117–34. New York: Springer-Verlag, 1979.

Woods, L. E. "Active Organic Matter Distribution in the Surface 15 Cm of Undisturbed and Cultivated Soil." *Biology and Fertility of Soils* 8 (1989): 271–78.

World Commission on Environment and Development. *Our Common Future.* Oxford: Oxford University Press, 1987.

Wright, Angus. "Rethinking the Circle of Poison: The Politics of Pesticide Poisoning among Mexican Farm Workers." *Latin American Perspectives* 13 (4) (1986): 26–59.

Yensen, S. B., and C. W. Weber. "Protein Quality of *Distichlis palmeri* Grain, a Saltgrass." *Nutrition Reports International* 35 (1987): 963–72.

Young, C. P. "Data Acquisition and Evaluation of Groundwater Pollution by Nitrates, Pesticides, and Disease-producing Bacteria." *Environmental Geology* 5 (1983): 11–18.

Young, Roberta A., and Gerald L. Horner. "Irrigated Agriculture and Mineralized Water." In *Agriculture and the Environment*, edited by Tim T. Phipps, Pierre R. Crosson, and Kent A. Price. Washington, D.C.: National Center for Food and Agriculture Policy, Resources for the Future, 1986.

Zadoks, J. C., and R. D. Schein. *Epidemiology and Plant Disease Management.* New York: Oxford University Press, 1979.

Zaki, Mahfouz H., Dennis Moran, and David Harris. "Pesticides in Groundwater: The Aldicarb Story in Suffolk County, N.Y." *American Journal of Public Health* 72 (1982): 1391–95.

Zaporozec, A. "Nitrate Concentrations under Irrigated Agriculture." *Environmental Geology* 5 (1983): 35–38.

Zimmer, D. E., and Dale Rehder. "Rust Resistance of Wild *Helianthus* Species of the North Central United States." *Phytopathology* 66 (1976): 208–11.

About the Authors

JUDITH D. SOULE grew up in Pullman, Washington, and earned her Ph.D. in plant ecology from Michigan State University under the direction of Dr. Patricia Werner in 1981. Dr. Soule conducted graduate research both at Kellogg Biological Station in Hickory Corners, Michigan, and at Rocky Mountain Biological Lab in Crested Butte, Colorado. She has taught at the University of Oklahoma Biological Station at Lake Texoma, served as research associate in ecology at The Land Institute in Salina, Kansas, and held various positions in The Nature Conservancy's Michigan Heritage Program, where she now serves as environmental review coordinator. Her current professional interest is in land use planning, which integrates sustainable agriculture with conservation of biodiversity. She lives in East Lansing, Michigan, with her husband, three children, two cats, ten bicycles, a quilt on a frame, and too many tomato plants to keep track of.

JOHN K. PIPER has been the ecologist at The Land Institute since 1985, where he teaches classes and directs studies of the prairie ecosystem and experimental polycultures of new perennial seed crops. He holds a degree in biology from Bates College and a Ph.D. in botany from Washington State University, and has taught at Washington State University and Benedictine College in Salina, Kansas. Mr. Piper and his family reside in Assaria, Kansas, in a farmhouse built in 1918.

Index

ALSO AVAILABLE FROM ISLAND PRESS

Ancient Forests of the Pacific Northwest
By Elliott A. Norse

Balancing on the Brink of Extinction: The Endangered Species Act and Lessons for the Future
Edited by Kathryn A. Kohm

Better Trout Habitat: A Guide to Stream Restoration and Management
By Christopher J. Hunter

Beyond 40 Percent: Record-Setting Recycling and Composting Programs
The Institute for Local Self-Reliance

The Challenge of Global Warming
Edited by Dean Edwin Abrahamson

Coastal Alert: Ecosystems, Energy, and Offshore Oil Drilling
By Dwight Holing

The Complete Guide to Environmental Careers
The CEIP Fund

Economics of Protected Areas
By John A. Dixon and Paul B. Sherman

Environmental Agenda for the Future
Edited by Robert Cahn

Environmental Disputes: Community Involvement in Conflict Resolution
By James E. Crowfoot and Julia M. Wondolleck

Forests and Forestry in China: Changing Patterns of Resource Development
By S. D. Richardson

The Global Citizen
By Donella Meadows

Hazardous Waste from Small Quantity Generators
By Seymour I. Schwartz and Wendy B. Pratt

Holistic Resource Management Workbook
By Allan Savory

In Praise of Nature
Edited and with Essays by Stephanie Mills

The Living Ocean: Understanding and Protecting Marine Biodiversity
By Boyce Thorne-Miller and John G. Catena

Natural Resources for the 21st Century
Edited by R. Neil Sampson and Dwight Hair

The New York Environment Book
By Eric A. Goldstein and Mark A. Izeman

Overtapped Oasis: Reform or Revolution for Western Water
By Marc Reisner and Sarah Bates

Permaculture: A Practical Guide for a Sustainable Future
By Bill Mollison

Plastics: America's Packaging Dilemma
By Nancy Wolf and Ellen Feldman

The Poisoned Well: New Strategies for Groundwater Protection
Edited by Eric Jorgensen

Race to Save the Tropics: Ecology and Economics for a Sustainable Future
Edited by Robert Goodland

Recycling and Incineration: Evaluating the Choices
By Richard A. Denison and John Ruston

Reforming the Forest Service
By Randal O'Toole

The Rising Tide: Global Warming and World Sea Levels
By Lynne T. Edgerton

Saving the Tropical Forests
By Judith Gradwohl and Russell Greenberg

Trees, Why Do You Wait?
By Richard Critchfield

War on Waste: Can America Win Its Battle with Garbage?
By Louis Blumberg and Robert Gottlieb

Western Water Made Simple
From *High Country News*

Wetland Creation and Restoration: The Status of the Science
Edited by Mary E. Kentula and Jon A. Kusler

Wildlife and Habitats in Managed Landscapes
Edited by Jon E. Rodick and Eric G. Bolen

For a complete catalog of Island Press publications, please write:
Island Press, Box 7, Covelo, CA 95428, or call: 1–800–828–1302.

Notes

Notes

Notes

Notes

Notes

Notes

Notes

Notes

Notes

Notes